THE "Countryman's Steam" MANUAL

John Haining AMIED

Nexus Special Interests

Nexus Special Interests Ltd.
Nexus House
Boundary Way
Hemel Hempstead
Hertfordshire HP2 7ST
England

First published by John Haining 1982
This edition published by Nexus Special Interests Ltd 1996

ISBN 1–85486–136–0

Phototypesetting by The Studio, Exeter
Printed and bound in Great Britain by Biddles Ltd., Guildford & King's Lynn

Contents

LIST OF PLATES

LIST OF FIGURES

Acknowledgements

I would like to acknowledge with thanks the assistance of Ted Jolliffe, E. Lancaster, D. Huw Jones and Earl Pennell in providing the photographs.

Dedicated to those model engineers whose constant support makes possible the Steam Corner at Usk Show.

Introduction

My first traction engine was built without a lathe, and with more than its fair share of soft solder, using a couple of planks in a dark cart shed as a workbench.

Despite a parental ban, the boiler was fired by a near-lethal small blowlamp, miraculously without damage to life or property.

By present-day standards that engine would be regarded with horror, but it worked, and it gained me an apprenticeship in steam when due to the depression, these were difficult to obtain. It also taught me a lot and gave, tinged with a spice of danger, a great deal of pleasure; today its chimney-base still reposes in my brass box, together with one hub and the nameplate.

The first *''Countrymans Steam'' Manual* was published in 1982 and was written around a few of the queries received from time to time from *Model Engineer* readers, together with some of the hints and tips which have proved useful to me over the years.

This latest Manual includes an extra chapter dealing with three different types of vertical boiler and has many additional and extended notes relating to welded steel boilers.

Perhaps a little of what is written may be of help to others, or may even introduce a newcomer to the most satisfying of all hobbies and that dignified and most lovable of beasts, the steam engine.

This book is dedicated to all who have so kindly expressed their appreciation of ''Countryman's Steam''.

John Haining
President, The Road Locomotive Society 1980–81
Tredunnock, Near Usk.

Foreword

Folklore and mystery surrounding the construction of small boilers has for too long bemused amateur and professional engineers alike.

John Haining has spent most of his working life designing and constructing steam engines and boilers. Who then, is better qualified to produce this manual for general use? Here he has described how to make and use small boilers in a variety of materials, with full backing of his experience in both theory and practice.

Tales of disasters with boilers are legion, but standards of safety today are such that only rarely – usually through ignorance or misuse – do these stories have any substance. Nevertheless, construction, testing, use and maintenance of a potentially dangerous boiler and its fittings must be to a high standard and this manual will be invaluable in maintaining and increasing safety standards.

My own introduction to steam power came comparatively late in life, so with others like me who did not have the benefit of a practical steam education – and perhaps some who have – I shall keep a copy of this book handy for guidance and commend it as a necessary part of every model engineer's (and many others) equipment.

Colin Tyler
Sutton Courtenay, Oxon.

Publisher's note
Sadly Colin Tyler died in 1995.

Plate 1 Caradoc nearing completion

Plate 2 Caradoc with Sentinel boiler

CHAPTER 1

Steel boilers and welding requirements

The high cost of copper tube and sheet has resulted in an increasing interest in welded steel boilers. Unfortunately, incorrect and distorted reports of material, testing, and insurance requirements – both at home and overseas – have not helped to present a fair case for consideration by model engineers contemplating using steel for the first time as boiler material.

Two boilers were designed for the 2 inch scale Durham and North Yorkshire traction engine, both of 5 inches o/dia, one copper and one steel, arc welded throughout (see Figs 1 & 2).

Many Durham engines are under construction, about half of them having steel boilers, several being built by professional engineers themselves engaged in the boiler and pressure vessel field of industry.

South Africa, for example, has the most stringent rules on pressure vessel design and construction but the Durham boiler, fabricated in unstamped hot-finished black mild steel has been accepted there without question.

The assembly of any form of pressure vessel by electric arc-welding requires a considerable degree of experience and skill in the use of welding equipment, and while the average model engineer's workshop is usually equipped to carry out brazing and silver-soldering techniques, comparatively few are able to meet the requirements laid down for the construction of welded steel boilers, particularly in the larger sizes. Consequently it is far better to allow this work to be undertaken by a firm specializing in steel fabrication, with the necessary techniques and equipment available to hand, than to attempt it oneself if doubtful of the result. Those model engineers lucky enough to be both competent welders themselves and in possession of adequate arc-welding equipment will be familiar with the informative booklets published by the British Standards Institution on the subject, in particular BS 2642 dealing with arc-welding of steel and BS 1500 and 2790 covering welded steel pressure vessels and boilers. These cover such subjects as electrode size relative to plate thickness, welding procedure, and types of welds, together with a great deal more information and recommendations.

A point to remember though, is that while BS 2790 Part 1. specifies requirements for boilers and vessels of various types, these apply only to those of 23.62 inches diameter and over, having a minimum shell thickness of ¼ inch and design pressure exceeding 105.5 psi. A further source of guidance is a small booklet published by the British Boiler Insurance Companies, which gives their recommendations for the design of steel pressure vessels, and their inspectors will usually require to see drawings and design calculations relating to a new boiler before proceeding with the hydraulic testing, to make sure that the design is basically sound and in line with accepted requirements.

The weld details tabulated at the end of this chapter formed the basis of a code of practice used by me in the design of pressure vessels in a variety of shapes and sizes; it is particularly useful in that it applies to plate thicknesses as low as ⅛ inch and can also be applied to boilers in 2 and 3 inch scale.

Preceding the details are some notes on boiler design and formulae.

The case for the arc-welded steel boiler rests on two absolutely essential requirements, the first being sound design and the second first-class fabrication and material; given these requirements in full, the small-scale boiler is practically indestructible at the pressures usually employed. The three principal enemies of any steel boiler are corrosion, neglect, and misuse; the first should be allowed for at the design stage, while the last two, under which should perhaps be included the damage possibly caused by the folly of "testing" to an excessively high pressure, can usually be ascribed to inexperience.

Selected to illustrate the construction of a typical steel boiler is that designed for the Durham and North Yorkshire traction engine in 2 inch scale.

A novel feature of the full-size engine, built at North Bridge Engine Works, Ripon in 1875 was the use of separate hornplates attached to the outer firebox wrapper plates, instead of being upward extensions of the latter as was the usual practice; thus the construction used on most models with copper boilers actually conformed to full-size practice in this case (see Fig. 3). The boiler barrel is a 16¼ inch length of hot-finish seamless mild steel tube, 5 inches outside dia and 3/16 inch thick.

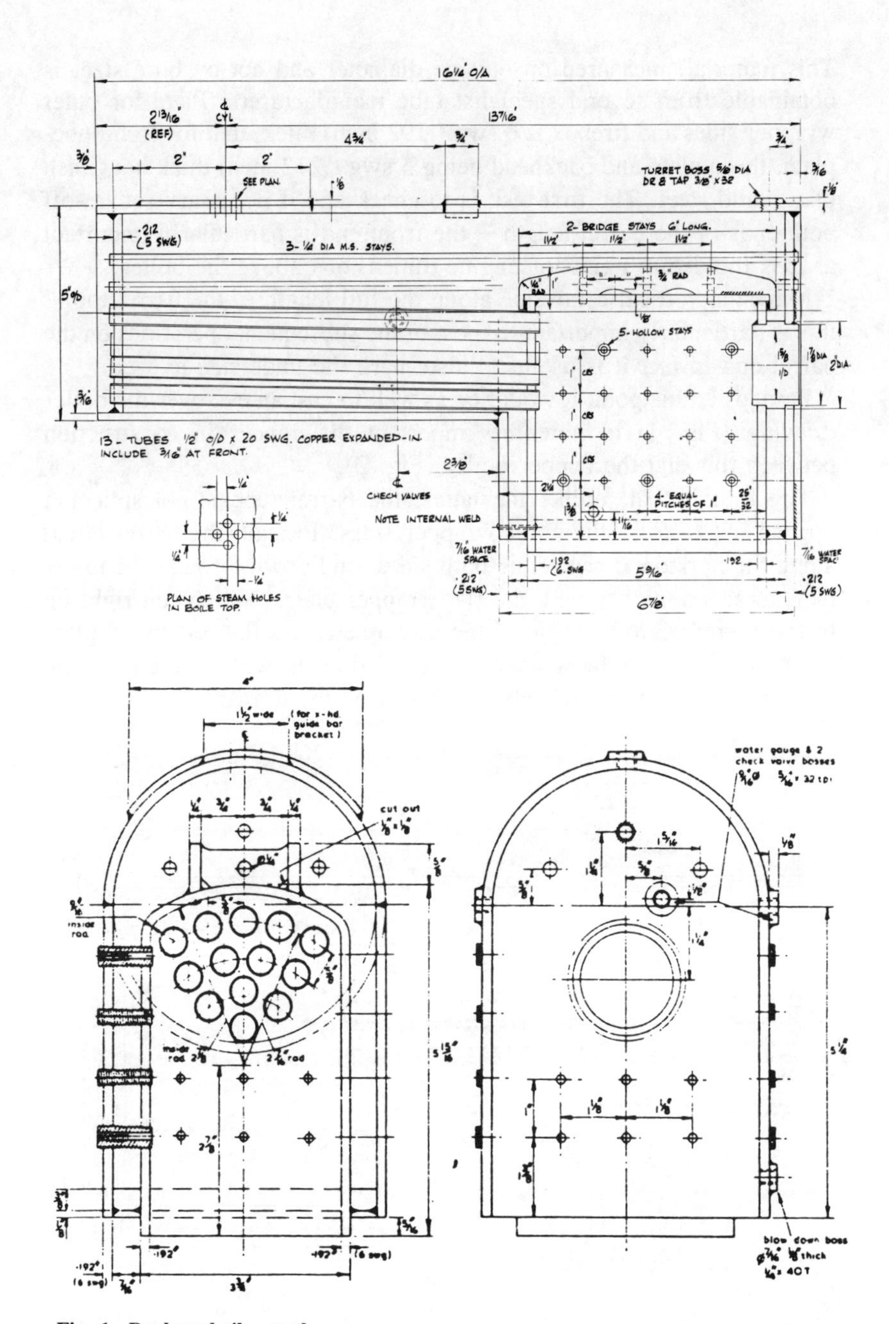

Fig. 1 Durham boiler steel

This material, measured on outside diameter and not by bore size, is obtainable from several specialist tube manufacturers. Plate for outer wrapper sides and firebox is 6 swg (.192 inch) thick, that for front tubeplate, throatplate and backhead being 5 swg (.212 inch) thick hot-finish black mild steel. The first task is to check and if necessary square-off both ends of the barrel length – the front end is particularly important as it is the datum point for engine dimensions above the boiler.

Mark the vertical centreline along the full length of the barrel top – this is particularly important to most of the subsequent operations on the barrel and in fact it is useful to also mark the underside as well.

Perhaps at this point it would be as well to cast an eye over the boiler drawing (Fig. 1) to note the important differences in construction between this and the copper boiler (Fig. 2).

The first is that, unlike the latter, the barrel tube is not split and opened out to form the outer wrapper sides. Instead, the barrel is cut along the horizontal centreline both sides, and down through the lower half, to accommodate both the flat wrapper plates, which run right up to the barrel centreline. Next, the throatplate is a flat unflanged plate which sits between the wrapper plates, following with its top edge the inner radius of the barrel tube as shown on the drawing.

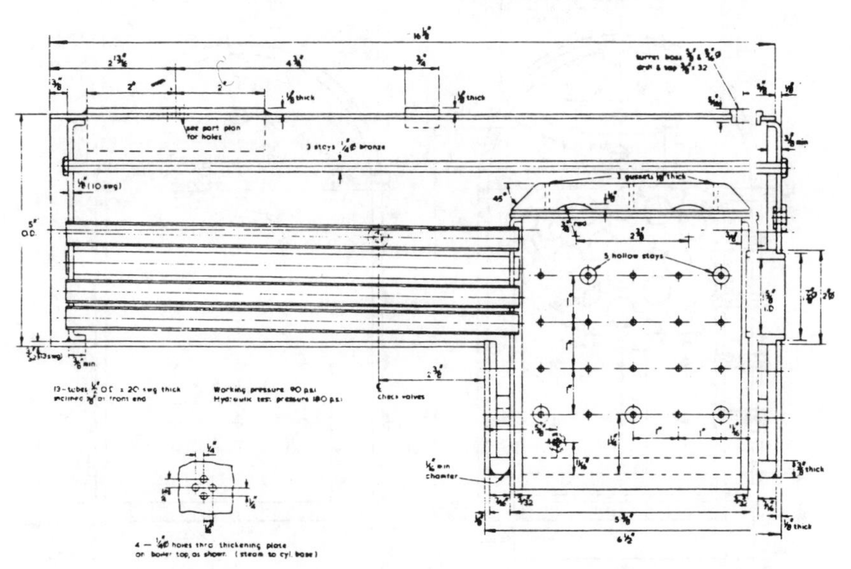

Fig. 1A Durham boiler in copper

10 - HOLLOW STAYS (5 - EACH SIDE)
3/8" x 32 T.P.I. TAPPED Nº 2 BA. 9/16" DEEP
42 - STAYS SCREWED 5/16" x 32 TPI.
ALL STAYS TO BE STEEL, TIGHTLY THREADED IN PLATES
LONGITUDINAL STAYS LIGHTLY RIVETED EACH END BEFORE WELDING.
BOILER TO BE M.S. THROUGHOUT, THE SHELL OF 5" O/D.
HOT-FINISH SEAMLESS TUBE. CUT TO RECEIVE OUTER WRAPPER PLATES
ALL WELDS CONTINUOUS-SEAM, HEAVY FILLET. ELECTRIC ARC
WORKING PRESSURE 85 PSI.
HYDRAULIC TEST PRESSURE 170 PSI.

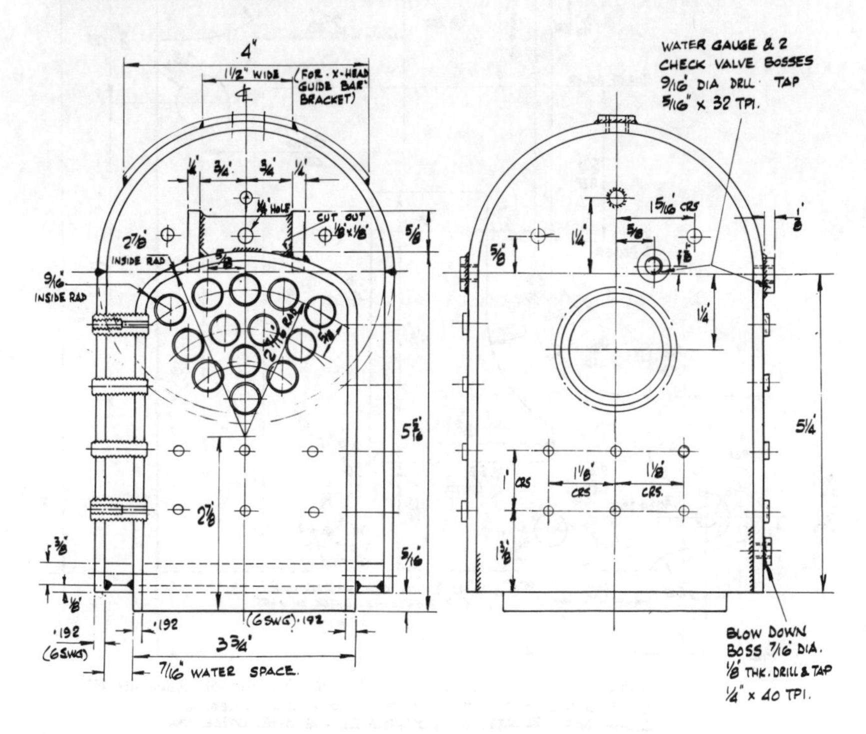

Fig. 2 Section through firebox etc. of Durham boiler

This is an important point to observe as the lower edge of the barrel supports the throatplate, and the weld, as well as forming a fillet on the outside radius, must be repeated around the inside to seal against corrosion forming at the junction of throatplate outer edge and inner edge of barrel.

Having cut out the barrel and made both wrapper plates, slightly chamfer the plate edges for the external welds, and when these are

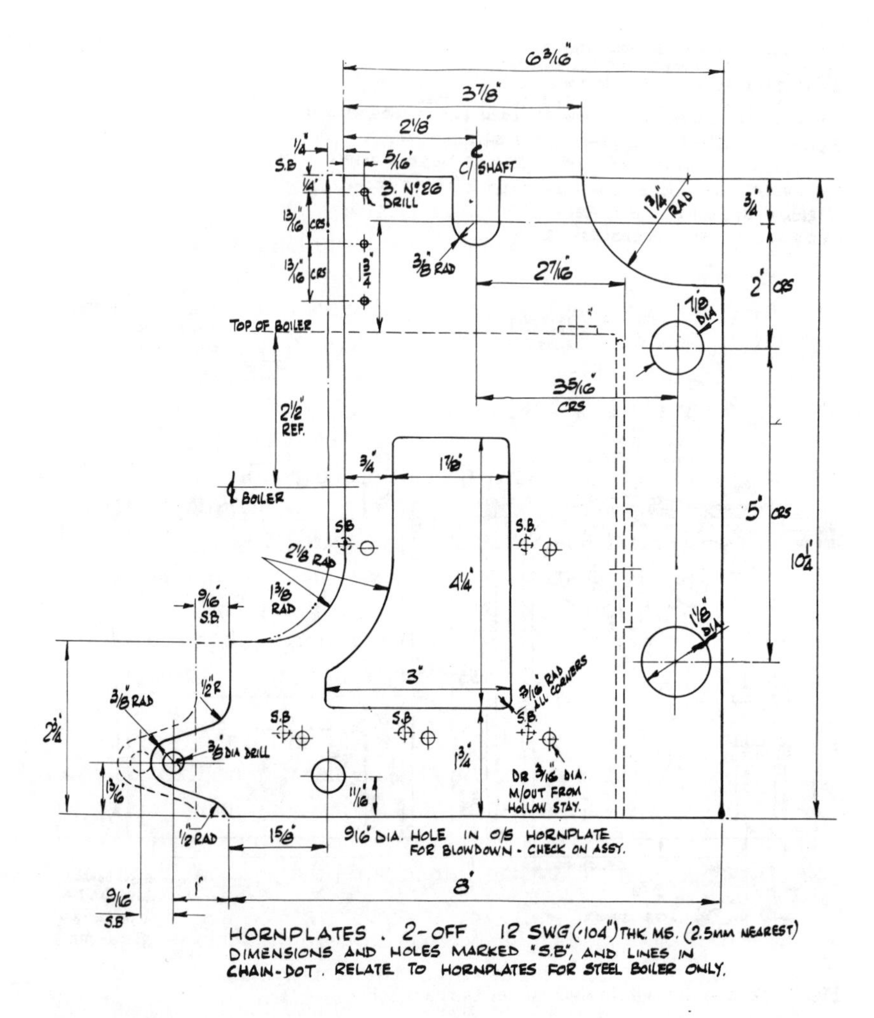

Fig. 3 Durham hornplates

completed, finish off with a sealing run inside between throatplate and wrappers.

Incidentally, the total length of cut-out section of the barrel will be 7 1/16 inches, as the backhead is inserted (3/16) inch inside the barrel to allow a good external fillet all the way round.

We should now have the barrel complete with outer wrapper plates and throatplate welded in position, and can go ahead with adding the various bushes, the cylinder reinforcing plate and the small rectangular pad for the crosshead guide bracket. I have long since abandoned the internally fitted cylinder reinforcing plate on both copper and steel boilers for this far easier method of going about the job.

All the bushes are turned from mild steel bar and unlike those on a copper boiler are not turned down to a reduced diameter to fit into the barrel or plate.

The front tubeplate can now be cut and turned to a neat fit in the barrel, with a slight chamfer on the outside edge in preparation for welding. Mark out and drill the tube holes slightly undersize to allow for finish reaming and bearing in mind that they are inclined 3/16 inch at the smokebox end, the bottom one lying just inside the plate weld. In the firebox tubeplate the second tube from the bottom on the boiler centreline is raised 1/16 inch, making it 9/16 inch from the row above, the tubes each side and the one below retaining their 5/8 inch pitch. Tubes should be a push fit in their respective holes, which will need the barest perceptible chamfer to remove burring. The drawing calls for copper tubes to be expanded in, but as alternatives they have been silver-soldered into front and firebox tubeplates and on a few boilers, have been of 18 swg steel lightly welded in place.

Expanded copper tubes possess an advantage in that the firebox can be built up and welded as a complete finished unit with bridge stays and firehole ring in place, then together with backhead and foundation ring, welded in position in the boiler. The tubes are then inserted and expanded as the final operation, following full-size practice – more about this later, though.

If either of the two alternative methods are used the tubes must all be silver-soldered or welded into the firebox tubeplate initially – the firebox being, of course, completed but not welded into place – in a similar fashion to building the copper boiler. The firebox is then placed in the boiler with tubes inserted into the front tubeplate, the foundation ring welded, together with backhead, and the operation completed by silver-soldering or welding all tubes into the front tubeplate. A word now regarding expanding the tubes. Use a taper drift in lieu of an expander ground to an included angle of not less than 2½ degrees and not more than 3½ degrees absolute maximum. Copper tubes should be soft enough to expand without annealing, but be very careful not to over-expand, or the tube will shear against the edge of the tubeplate hole.

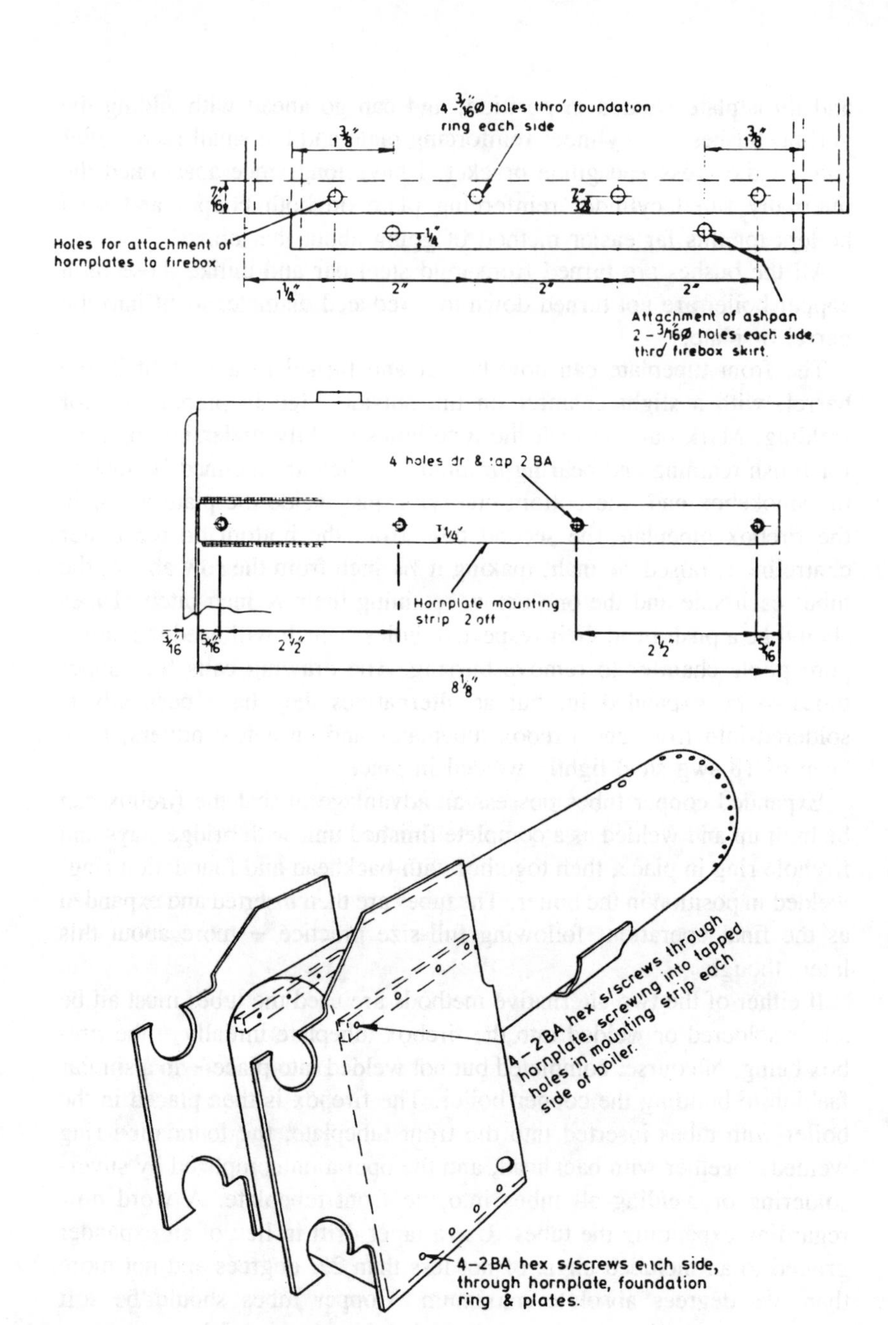

Fig. 4 Attaching hornplates to firebox

There is just enough space in the firebox to handle the drift, and the big firehole is a considerable help. Dealing with the firebox itself, there are several schools of thought on how best to bend the rather heavy gauge firebox wrapper. If oxy-acetylene is available my own preference is to use two ½ inch thick mild steel formers the same size as the firebox endplates, but left over-length at the bottom for holding in a vice. Weld a couple of steel bars or lengths of pipe, turned down to give the corner radii, between the plates, with another length of pipe or bar lower down to keep the job square.

Cut the wrapper plate, allowing at least ¼ inch on the width and a few inches on the developed length; grip the former fixture in the vice with the wrapper held tightly between the two formers and the jaw and heat along the first corner radius line to a dull red, bending the plate over the formers and repeating the heating process to form the crown and opposite corner radii. Draw the wrapper off the formers and leave to cool before trimming the lower edge to finished length. Cut out the firebox tubeplate and backplate, using the steel formers as a template and cleaning up the edges to a neat fit inside the finished wrapper.

Clamp the wrapper over the front and back plates with a block of hardwood between them to maintain the correct distance, and not forgetting to leave an equal overlap outside both plates for the external seam welds. Follow by seam welding along the inside (after removing wood block!), using No 10 rod to give a good continuous seam and medium heavy fillet.

Test all seams by inverting the firebox with about ¼ pint of paraffin inside and tilting to cover each seam in turn. If traces show on the outside, empty completely and after grinding-out, re-weld the faulty area, before welding the bridge stays to the crown. Holes for tubes and the firehole can be drilled at this stage, the latter marked off from the backhead. The firehole tube can be inserted into the firebox back plate and welded, the tubes dealt with as already described, and after all surplus plate has been ground or trimmed off, the completed box can be dropped into the inverted boiler for the final tasks of welding the foundation ring and backhead. The foundation ring is made up of 4 lengths of ⅜ inch thick × 7/16 inch wide ms bar, well chamfered on the underside for welding. It is important that these are a really tight fit all the way round to avoid potential corrosion traps between plates and bars. Sometimes these are welded to the firebox along the top of each bar, before inserting the box into boiler, but this still does not seal between foundation ring and outer plates. With the firebox in position in the boiler, insert the front length of foundation ring and secure to firebox and throatplate with a couple of 3/32 inch dia iron rivets. Next, fit the backhead over the firehole ring, the front tubeplate into the barrel (if this has not already been done), and insert the 3 longitudinal stays through both, leaving over-length for cutting off after welding. Fit the two side and the back length of foundation ring, securing each with 2

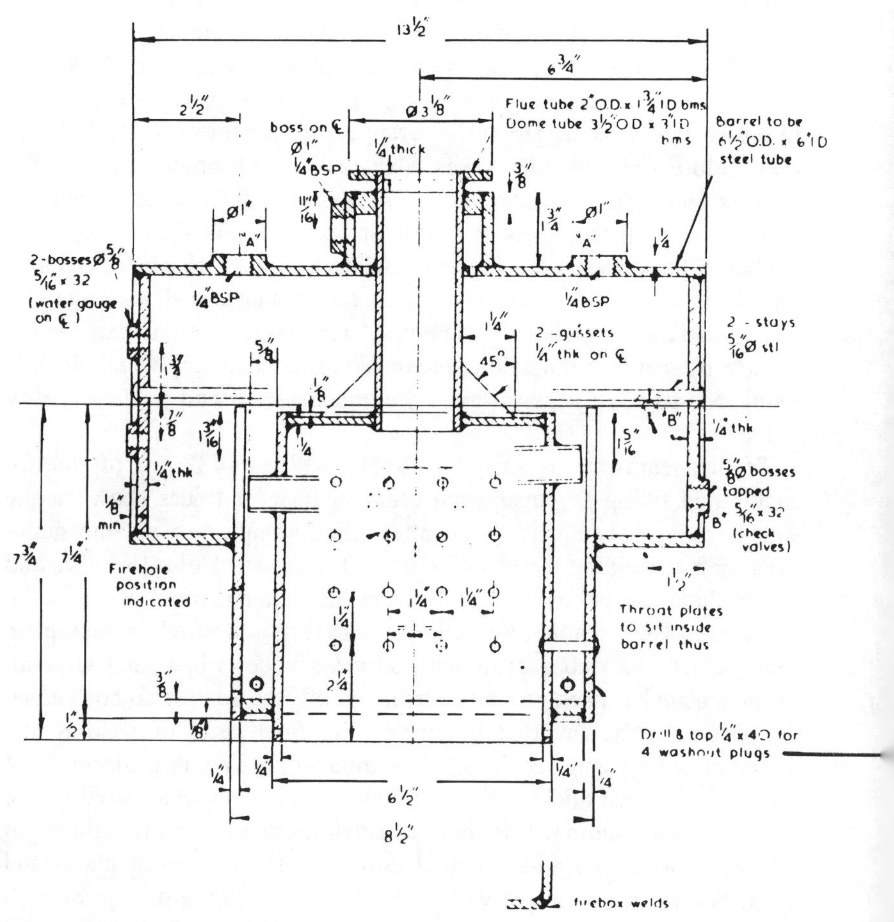

Fig. 5 Suffolk boiler sections

rivets, weld up the backhead followed by the foundation ring complete, and after making sure the two sets of tube holes line-up (if tubes are not in position) and are square to the boiler centreline, weld in front tube-plate and cut off the projecting spare stay lengths, welding over each stay where it sticks through the tubeplate and backhead. Now drill and tap for all the firebox stays – a cardboard template helps here – leaving threads tight and smeared with a spot of Loctite before screwing home

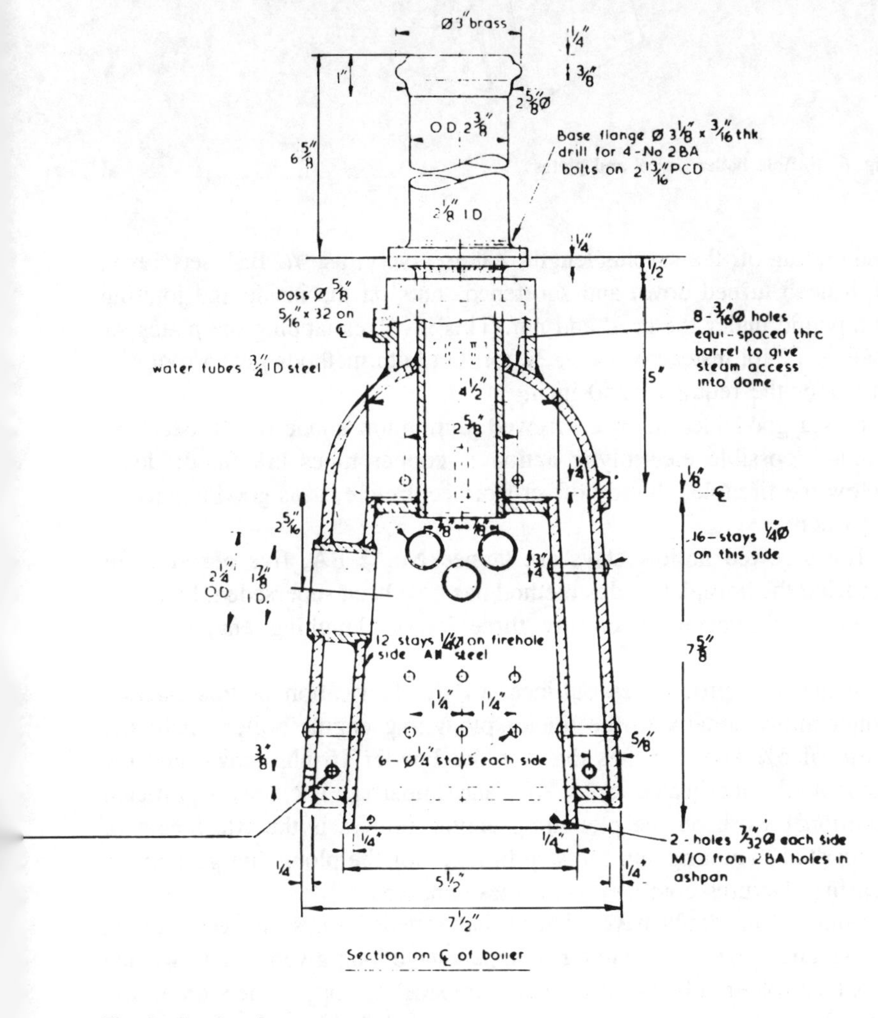

Fig. 5 Continued

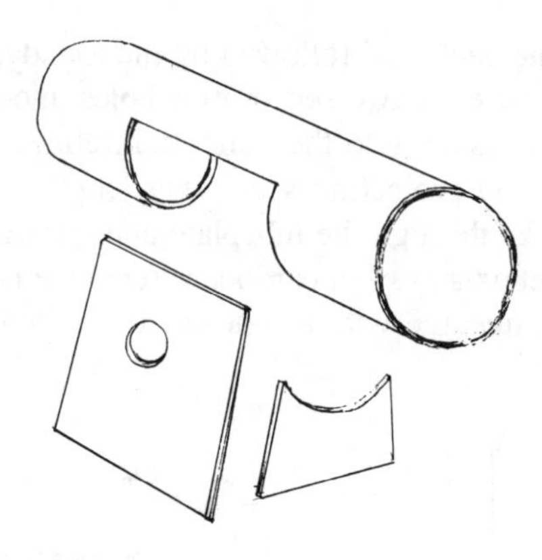

Fig. 6 Suffolk boiler barrel and plates

and cutting off the surplus length. Alternatively, use 5/16 BSF setscrews, with head turned down and shortened, nuts on the inside and jointing compound under the head and nut. This involves tapping the plates 5/16 BSF × 22 tpi instead of 5/16 × 32 tpi, but both methods give a root dia. just over the required .250 inch.

It is a good idea to fit a screwed aluminium anode or "waster" to counter possible electrolytic action if copper tubes are fitted; down below the firehole, slightly offset from centreline, is as good a place as anywhere.

The screwed hollow stays are tapped No. 2 BA, five per side for securing the hornplates; this method has now been superseded, by a new method of attachment on my three latest ploughing engine boiler designs. (Fig. 4)

In general, procedures outlined for the fabrication of this traction boiler apply equally to the longer ploughing engine boiler, including those of 5¾ and 6 inches dia. for the Fowler 16 nhp single and the compound superheated class Z7S and Superba. The only significant additional work on the ploughing engine boilers is the attachment of drum stud mounting and footstep bracket for the ploughing gear lower bearing, features common to all these engines.

You will no doubt have noticed that screwed stays between firebox crown and boiler top (or outer wrapper) have long given way to welded or silver-soldered bridge stays cut from steel or copper sheet on all my boilers. These have cut-out apertures and holes to ensure water circulation and to assist in preventing build up of sediment around the firebox crown. The radiused form of firebox top possesses sufficient inherent strength to withstand normal pressures, even without the welded and linked bridge stays, but these should still be included, while the foundation ring welds also support the firebox against downward pressure on the firebox crown.

For many years it was argued that firebox crowns must be supported by screwed rod stays to the outer wrapper or boiler top, and that inadequate staying imposed excessive loading on foundation ring rivets. Many early fireboxes were flat topped and so required staying to the boiler top for support, and it was not until the corrugated or arched firebox top was successfully introduced and adopted by a number of makers that the argument was proved to be false.

Aveling and Porter used a box with double corrugated crown, and the Marshall S Series Roller boiler had a very neat cruciform corrugation or indentation in the top of the firebox, to name just a couple of variations employed.

In common with a few other early engines, the Durham and N. Yorks traction engine had no manhole or McNeil ring giving access to the boiler, and even had one been fitted in full size, to reproduce in 2 inch scale seemed to invite possible weakening of the shell for no real advantage. Mud plugs were commonly fitted at each corner of the firebox, just above the foundation ring, to enable each length of lower water space to be hosed out; my drawing calls for two ¼ × 40 tpi screwed plugs low down on the backhead for this purpose, more necessary on the steel boiler than on the copper version.

The Suffolk boiler

This steel boiler was developed from one designed to retain the external lines of a Yorkshire steam wagon and tractor boiler while at the same time combining ease of construction with good steaming qualities. It is not a true return tube boiler, as is the original Yorkshire.

The first one was fitted in my Suffolk dredging tractor, from whence it derives its title, the identical boiler having been constructed by readers of *Model Engineer* building this engine from my constructional articles, while a larger version has been constructed for use in a boat.

Mr. Colin H. MacDonald from Scotland has introduced a neat and very useful modification in that both endplates of his barrel have been made detachable and secured to machined flange facings each end by 20 setscrews thus allowing entry to the boiler for cleaning and inspection. The longitudinal stays are in the form of long bolts and nuts, with special chamfered washers and PTFE grommets sealing each end.

The standard boiler is a very free steamer and my own is used not only in the Suffolk tractor but as a workshop boiler and even, on occasion, to sterilize seed beds.

A steam dome around the chimney base overcomes the bogey of all transverse boilers on rough ground – the liability to prime and keeps steam dry enough to dispense with the need for a superheater.

Four washout plugs just above the foundation ring help to keep the water spaces free of sediment, while a high pressure jet inserted through the two top screwed bosses will wash out above the crown plate. Extension of all tube ends outside the firebox plates helps to reduce what might have remained as deadwater areas at each end of the horizontal barrel. The angular inclination of the tubes together with generous water spaces around the firebox have a great deal to do with the good steaming properties of this boiler. Fig. 5 shows two sectional views of the boiler and Fig. 6 the boiler barrel cut for wrapper and throatplates.

Services of a coded welder

To ensure first-class fabrication, it is essential to entrust the welding to an experienced and certificated welder. Most will accept a boiler for welding which has been assembled or partly-assembled, with plates clamped or otherwise held in place. Some will accept a boiler in this state with any necessary plate preparation already carried out while others may prefer to carry out all preparation themselves. In the matter of plate preparation, some welders may have ideas which differ slightly from the laid-down requirements but result in perfectly sound welds which pass inspection. More of this later. Before looking at the actual construction of a couple of steel boilers, it may be of some interest to consider the characteristics of an arc-welded joint.

Welding methods compared

Arc welding has been defined, somewhat obviously, as a fusion process in which an electric arc is struck between the base metal and a melting electrode, fusing the adjacent edge of the base metal and supplying the filler metal. The actual weld itself is described simply as a localized consolidation of metals. The *efficiency* of either an arc-welded or an oxy-acetylene welded joint is the ratio of the strength of the welded joint to that of the original plate (the base metal).

This can be obtained by two methods, the first being a comparison of the strength of a test piece containing the welded joint with the strength of a test piece of equal width cut from the original plate, with no allowance for any additional thickness of the joint due to the addition of filler material in the welder. The second method of obtaining the efficiency of the joint material is based on the intensity strength at yield point of both joint and plate as calculated from the load and cross-sectional dimensions of both, i.e. load over area for mild steel plate based on a maximum permissable stress of 14,500 lb per sq inch at a temperature of up to 650 degrees F, or a pressure up to 6300 lb per sq inch at 900 degrees F.

Tests made on arc-welded joints in steel plates (about 0.6 per cent of carbon) showed that the ultimate strength of small welded specimens was over 100 per cent of the strength of the original plate for the thickness up to ½ inch but reduced to around 90 per cent for ¾ inch to 1 inch thick plate – the latter result being hardly likely to concern the average model engineer contemplating construction of a steel boiler, however.

Butt welds tested gave a tensile strength of from 90 to 96 per cent, while lap welds with double fillets (on both edges) gave a tensile strength of from 70 to 80 per cent of the original plate.

Tests showed that oxy-acetylene welds were considerably weaker under shock loading than the original plate material and for this and other reasons were generally inadmissible for pressure vessel fabrication.

Materials and preparation

The materials used in my steel boilers are hot-finish seamless mild steel tube and hot-finish black mild steel plate of medium to low carbon content. The supplies available nowadays are consistent in quality, but most suppliers are usually willing to give customers a specification of the plates they offer, should it be required.

WELD DETAILS.

FLAT BUTT WELD.

WELD E, F & G TO BE SEALED AT BACK.

SYMBOL	EXAMPLE.	WELD.	PARTICULARS.	
A.		LIGHT	L = 2T. D = $\frac{T}{2}$	FOR $\frac{1}{8}$" PLATES.
B.		HEAVY	S = $\frac{T}{2}$.	
C.		LIGHT	L = 1·5T. D = $\frac{T}{4}$	FOR $\frac{3}{16}$"; $\frac{1}{4}$"; $\frac{5}{16}$". PLATES.
D.		HEAVY	S = $\frac{T}{2}$	
E.		LIGHT	L = T	FOR $\frac{3}{8}$"; $\frac{7}{16}$"; $\frac{1}{2}$". PLATES.
F.		HEAVY	L = 1·5T. D = $\frac{T}{4}$.	
G.		LIGHT	L = 1·5T.	FOR $\frac{9}{16}$" PLATE & OVER.
H.		HEAVY	L = T. D = $\frac{T}{8}$	

LAP & FILLET WELD.

R.		LIGHT	L = T. O = NOT LESS THAN T.
S.		HEAVY	L = 1·4T. O = NOT LESS THAN T.
T.		LIGHT	L = $\frac{T}{2}$. $\frac{3}{16}$" MIN.
U.		HEAVY	L = 1·4T.

Corner Weld.

Symbol	Example.	Weld	Particulars.
L.	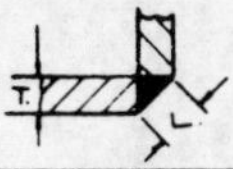	Light.	L = 1·4T.
LR.		Full strength.	L = T.
M.	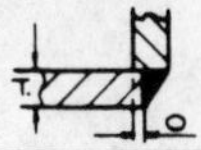	Light.	O = ·3T.
MR.		Full strength.	L = T. O = ·3T.
W.	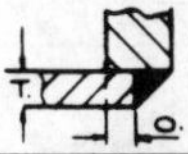	Light.	O = Difference in plate thickness.
WR.		Full strength.	L = T. O = Difference in plate thickness.

W & WR are for unequal plate thickness.

Tee Weld.

J.	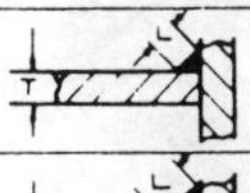	Light.	L = $\frac{T}{2}$; $\frac{3}{16}$″ min. single or double weld to be specified.
K.	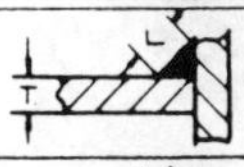	Heavy	L = 1·2T. or use double 'J'
		Full strength	L = 1·2T. double weld.
N.		Full strength	L = 1·15T. C = T for $\frac{3}{8}$″ to $\frac{5}{8}$″ plate, when double weld cannot be used.
P.	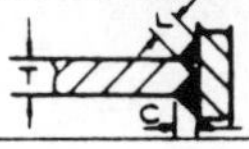	Full strength	L = ·6T. C = ·5T. for plate above $\frac{5}{8}$″ when weld 'K' cannot be used.

Plate 3 2 in. scale Fowler boiler

Plate 4 2 in. scale Fowler backhead

Plate 5 Merryweather Valiant in steam

No DIY welding

In describing the construction of a typical 2 inch scale traction engine boiler in steel, I have covered each stage of the assembly as if we were first putting the pieces together and then welding them in place, but in fact, unless you are a first class certified welder and competent to fabricate pressure vessels, do ***not*** attempt the welding or your boiler. I have made this point emphatically in other books but it is so important that it will bear repeating. The welding of any pressure vessel is of such vital importance that it should only be carried out by an experienced certificated welder, and the preparation of the component plates and tubes should be to his requirements and approval. To this end, the boiler can be prepared by cutting and fitting together all the platework, but all parts should be either clamped or held together by small angle brackets or plates, if necessary drilling and tapping the smallest possible holes for any necessary drilling setscrews. *Never* spotweld any of the component plates together, as all such spotwelds will have to be removed by grinding out before the welding seams are started and the welder will probably insist that this is done, even though a lot of time and effort will be wasted. Any small tapped holes can easily be filled up and welded over and most welders will not object to this, neither will most of them object to welding up the odd test piece or so, if you wish to check the calculated strength of a weld.

CHAPTER 2

Steel boilers – vertical types

Sentinel pattern boiler

Probably one of the most efficient cross-water tube boilers ever designed was the Sentinel. The boiler was first produced in 1915 for the original standard Sentinel waggons and remained in production in all its varying forms until steam waggon production ceased at Shrewsbury in 1950. The outer shell of the boiler was flanged inwards at the top and outwards at the bottom to mate with the outwards formed flanges of the tubular firebox. The firebox middle section was pressed into a square in plan view to form four flat tubeplates between which two sets of inclined water tubes passed, the four layers of tubes in alternate rows lying at right angles with space in the middle through which passed the vertical firing chute. A superheater coil surrounded the firing chute in the upper smaller diameter circular length of the firebox which was concentric with the larger diameter lower length.

The boiler shown in Fig. 7 was designed to power a compound Vee engine and it was intended from the outset that the boiler design should follow the Sentinel layout with top firing and slightly inclined cross-water tubes and flanged withdrawable firebox.

A necessary and regretted departure from the full-size Sentinel boiler design is that the firebox is not tapered upward and is not a single pressing. The Sentinel pattern of firebox has been followed in that the shape runs from circular to square to circular, allowing the tubes to be fitted into four faces and the firebox can be withdrawn as a complete unit from the shell.

The boiler barrel is made up from a single length of 15 0/dia hot finished seamless steel tube, minimum thickness ¼ inch. It is essential that both the ⅝ inch thick flanges at top and bottom are set absolutely square to the centreline and inner diameter of the shell to ensure an even

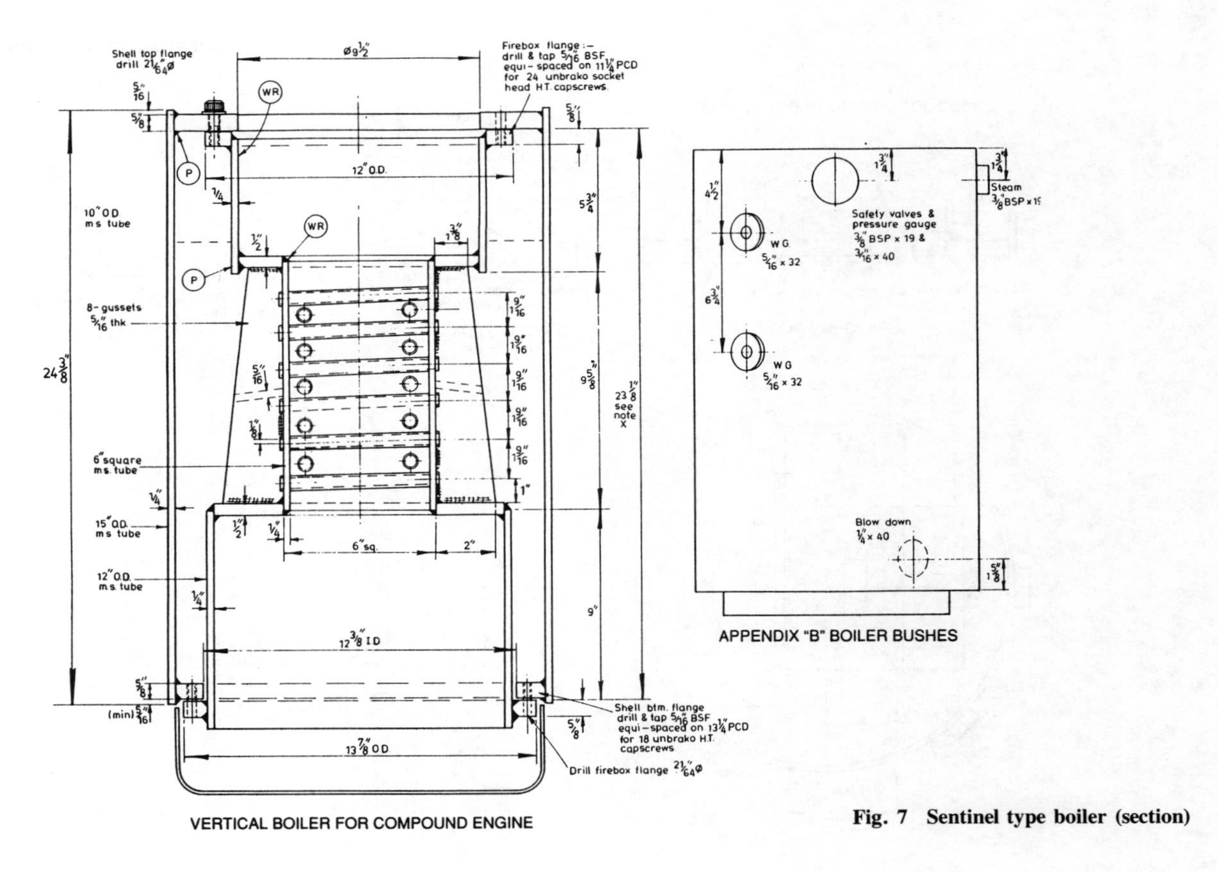

Fig. 7 Sentinel type boiler (section)

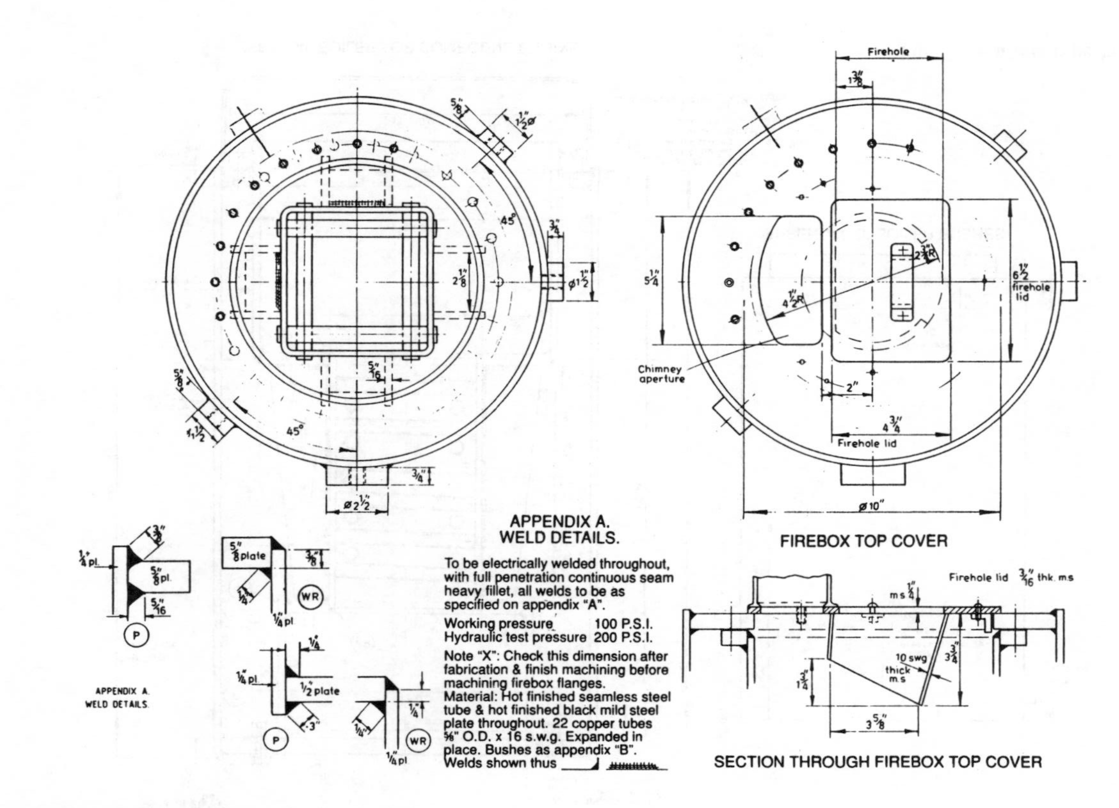

Fig. 8 Sentinel type boiler (plan)

thickness after machining.

The bottom flange ring and the top one of the firebox must, like the shell flanges be absolutely square to the axis of the firebox and the i/dia of top and bottom circular sections.

All flanges on finish machining must pull down to form a metal to metal joint with Red Hermetite as jointing.

Unbrako socket head screws with their high tensile and shear strength are used for both top and bottom flanges.

The material used in pressing (in one single heating) all Sentinel fireboxes was mild steel having a carbon content of 0.015 to 0.25 or 0.25 to 0.5 per cent and low to medium carbon black mild steel is specified for all my boilers.

Caradoc boiler

Two boilers were designed for the 3 inch scale steam tractor Caradoc.

The first was a multi-tubular boiler with 21 copper tubes 9/16 inch 0/dia expanded in, and a submerged top tube with space for a single turn of the main steam pipe to give a mild degree of superheat. The barrel is a 14 inch length of 8½ 0/dia hot finished black mild steel tube, with a circular firebox of 6 inch 0/dia tube ¼ inch thick and requiring no side stays. The second boiler was designed as an alternative for those builders who prefer top firing of 8¾ 0/dia with a centre flue of 4½ inch 0/dia steel steel and a firebox of 6½ 0/dia steel tube. 4 steel cross tubes are inclined at 2½ degrees arranged in vee formation in plan view. These tubes can be arranged to lie parallel across the flue at each side so as to allow space for dropping the coal down on to the fire as an alternative to the vee layout. A rocking grate and hinged ashpan facilitate clearing the firebox.

Unlike the first boiler, the chimney is offset forward to give as much of a Sentinel appearance as possible. My own Caradoc is fitted with this boiler. Working pressure is 95 to 100 psi on both boilers but this pressure is rarely needed for full performance.

Boiler tubes

The roller-type tube expander is now obtainable from trade sources to fit the tubes of small boilers.

As mentioned earlier a drift may be used if ground to the correct angle

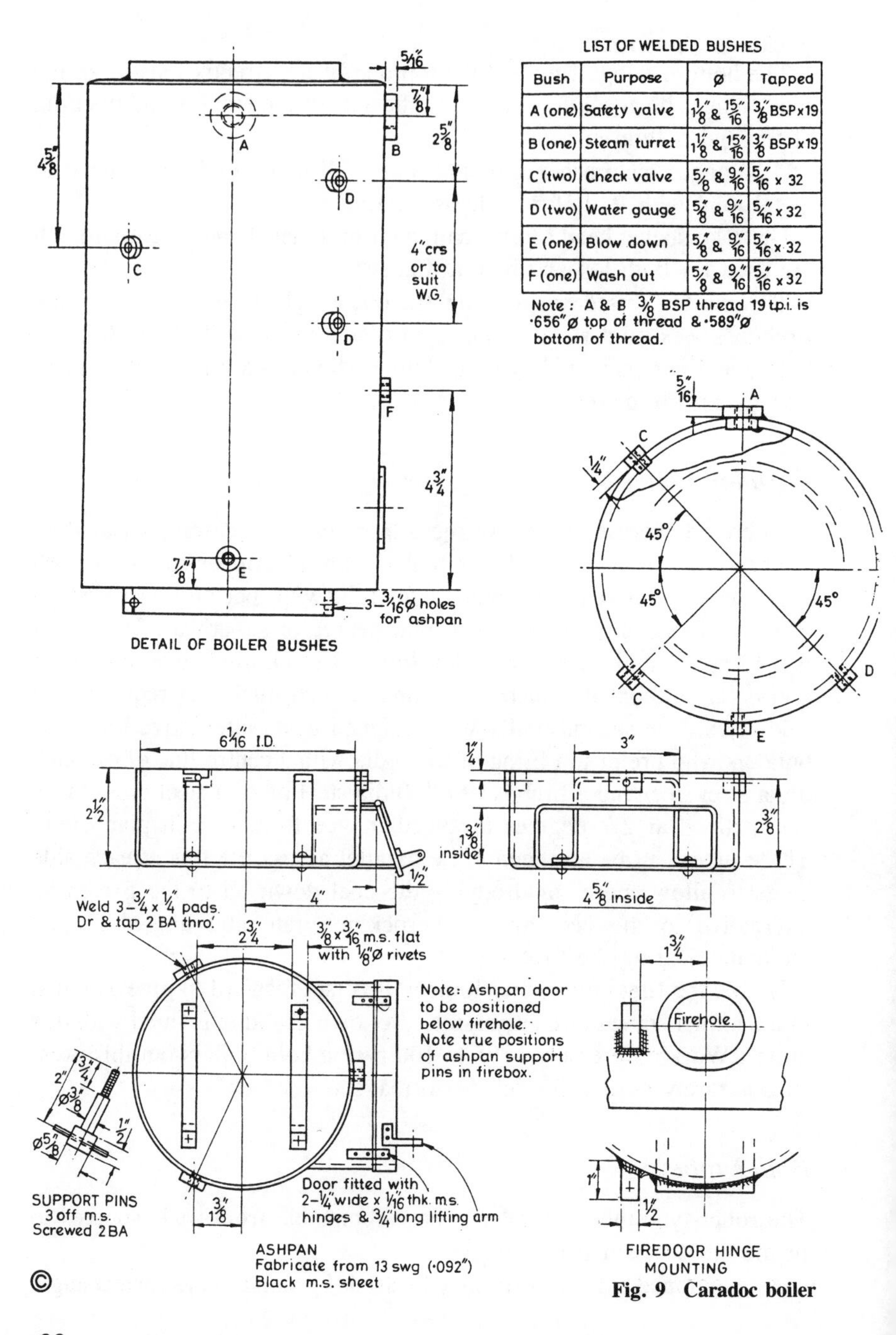

LIST OF WELDED BUSHES

Bush	Purpose	ø	Tapped
A (one)	Safety valve	1 1/8" & 15/16"	3/8" BSP x 19
B (one)	Steam turret	1 1/8" & 15/16"	3/8" BSP x 19
C (two)	Check valve	5/8" & 9/16"	5/16" x 32
D (two)	Water gauge	5/8" & 9/16"	5/16" x 32
E (one)	Blow down	5/8" & 9/16"	5/16" x 32
F (one)	Wash out	5/8" & 9/16"	5/16" x 32

Note: A & B 3/8" BSP thread 19 t.p.i. is ·656"ø top of thread & ·589"ø bottom of thread.

Fig. 9 Caradoc boiler

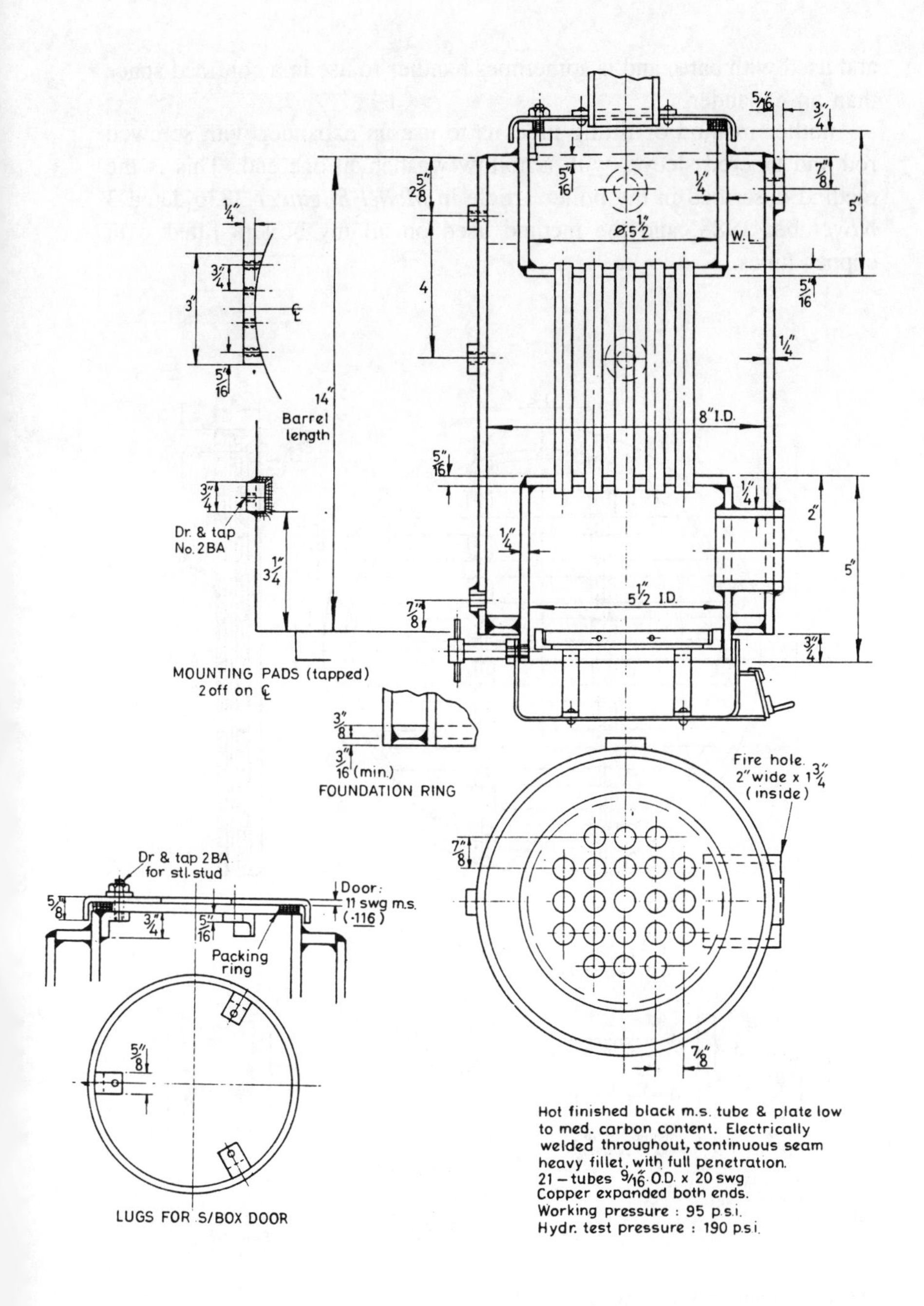

Hot finished black m.s. tube & plate low to med. carbon content. Electrically welded throughout, continuous seam heavy fillet, with full penetration.
21 – tubes 9/16" O.D. x 20 swg
Copper expanded both ends.
Working pressure : 95 p.s.i.
Hydr. test pressure : 190 p.s.i.

and used with care, and is sometimes handier to use in a confined space than an expander.

Another method of fitting tubes is to use an expander with screwed rod and tapered sleeve, with a hollow washer on one end. This is the method described in my boiler article in *Model Engineer* 3836 dated 3 November 1988, and the method used on all my boilers fitted with copper tubes.

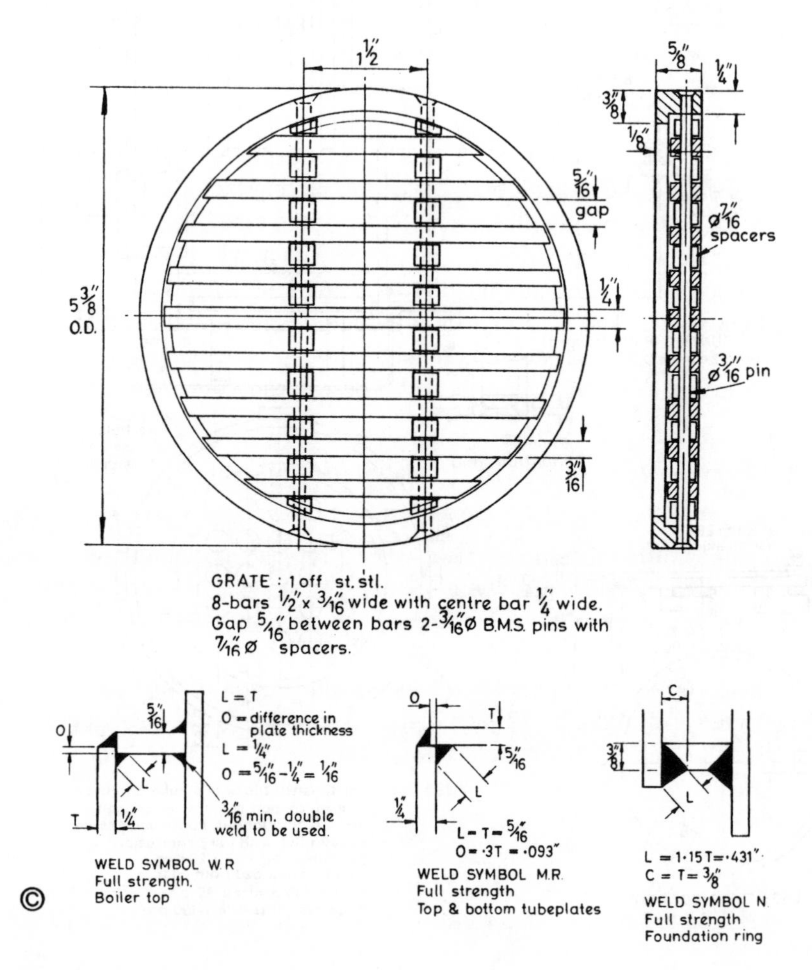

Fig. 10 Caradoc boiler's grate and welds

I believe the original pattern, as devised by A.A.Leak, was made as a double-ended tool, but it is easier with the rod expander to expand only one end at a time, owing to the danger of tubes moving lengthways or even dropping down inside one tubeplate when trying to eject the tapered collar wedged inside a tube. If a recessed washer is placed over the opposite tube end and both ends of the screwed rod reduced in diameter for a length of about ¼ inch to just below thread root diameter as shown on the drawing, the tube can be expanded nice and evenly by tightening the hexagon nut at the tapered sleeve end, the recessed washer at the other end taking the load against the other tubeplate. If a sharp tap on the reduced diameter nose of the screwed rod fails to release the tapered sleeve inside the expanded tube end or the tube shows signs of moving, take out the long screwed rod altogether and replace with a ¼ inch BSP or even No. 2BA screwed bar with a head large enough to sit below the tapered sleeve and using a bridge piece straddling the tube end, tighten the hexagon nut to draw the assembly out of the tube. This ensures that the taper sleeve does not drag the tube back with it. The headed extractor can of course be made any length to suit the tube length. These modifications to the original design of this rod do help to prevent that most infuriating happening in assembling a boiler – a tube moving outwards with a locked expander after one end has been expanded into the tubeplate.

A Merryweather boiler in 3 inch scale (Fig. 11)

Merryweather was particularly famous for its Valiant vertical boilered fire pumps, often mounted on two wheels for easy portability. As well as fire engines the firm also built specialist steam vehicles such as gully-suckers etc. using a larger version of the 26¾ inch 0/dia fire engine boiler.

The boiler described here is a 3 inch scale version of the fire engine, one at 6¾ inch 0/dia by 13 inch high. Like the Sentinel boiler, the firebox can be removed completely for maintenance.

Exactly ¼ of the 26¾ in 0/dia full-size boiler gives us 6.687 inch 0/dia so I settled for 6¾ inch 0/dia which is not possible to obtain (as far as I can find out) in steel tube form, the nearest being 6½ inch 0/diameter. This leaves us with the need to follow full-size practice exactly and roll up the shell, with a welded vertical seam and welded butt strap. A slight departure from the full-size design is in the boiler top plate, which instead of being a flanged pressing, riveted inside the boiler shell, is a flat plate inserted for a third of its 0.212 inch thickness into the shell and welded externally and internally to form an approved BSS joint for a plate as fully supported as this one is.

Still at the top of the boiler, the supporting angle ring is thickened up considerably to give more depth of thread and in addition, I would prefer to see the studs (2 BA) screwed right through the ring and nutted onto the underside. At the lower end of the parallel length of boiler shell, the plate should be slightly set back inside the steel angle ring to accommodate a sealing weld and the same applies to the angle ring welded to the top of the outer firebox skirt plate. The angle rings are drilled 3/16 inch dia for 2 BA steel hex. bolts, the top angle ring only having studs. More on these later. The safety valve boss and stop valve boss are welded into the boiler top plate, both tapped to suit their fittings; bosses are 3/4 inch and 5/8 inch dia suggested tapped 3/8 inch × 32 tpi. Water gauge bosses and blowdown boss are shown, to be tapped to suit their fittings. A third, smaller boss should be added, for the pressure gauge connection and syphon pipe, preferably on the boiler top.

The top angle ring is welded, as shown, to the 3 inch 0/dia. flue, which itself is welded to and projects down through the firebox crown plate.

I have shown the parallel length of the firebox flared outwards to fit over the tapered skirt of the firebox. This should be welded inside and outside as shown on the drawing. A point about this seam in the firebox is that the flare can be eliminated and the parallel length of plate welded directly to the top of the tapered skirt as an easier alternative, but it is advisable to obtain the approval of your assurer while retaining the same angle of inclination as in the full-size boiler for the cross-water tubes. Many full-size vertical boilers, usually steam waggon boilers, used a flat face pressed in the firebox sides, or even a square section, to facilitate

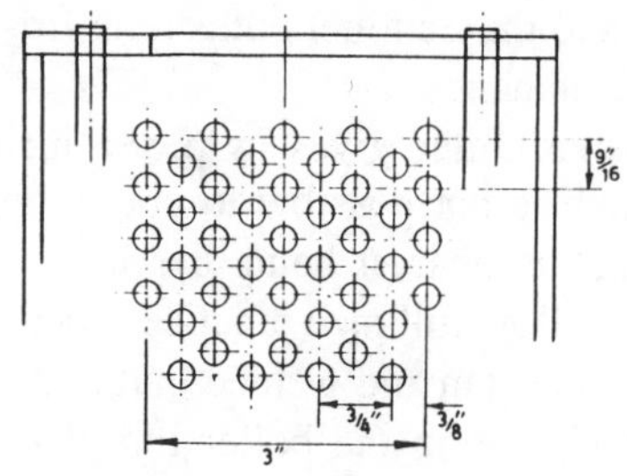

LAYOUT OF 43 CROSS TUBES

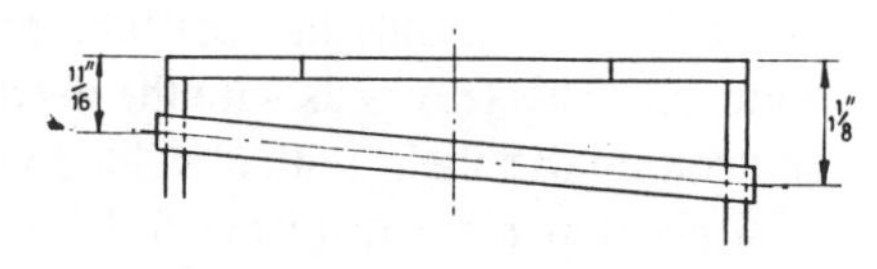

ALL CROSS TUBES INCLINED THUS

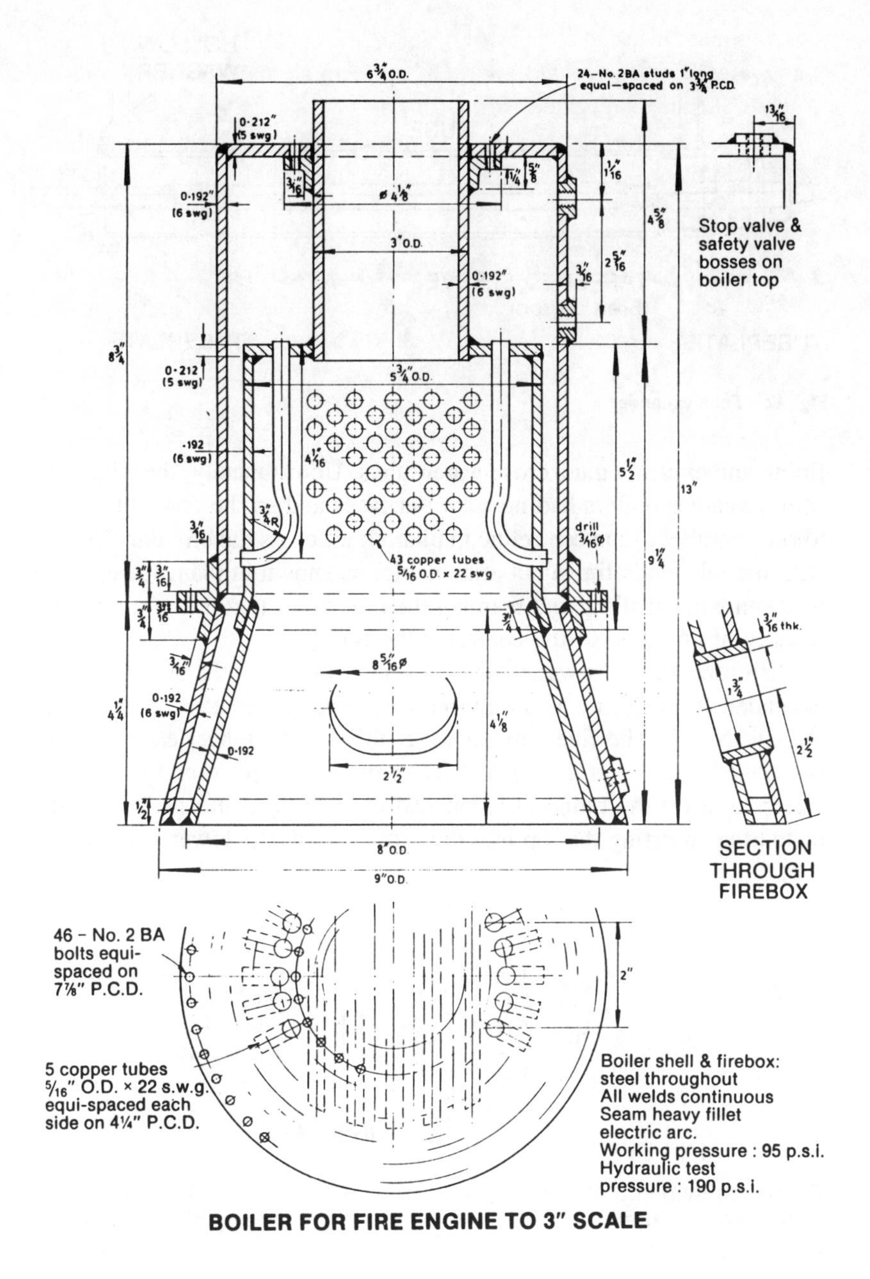

Fig. 11 Merryweather boiler

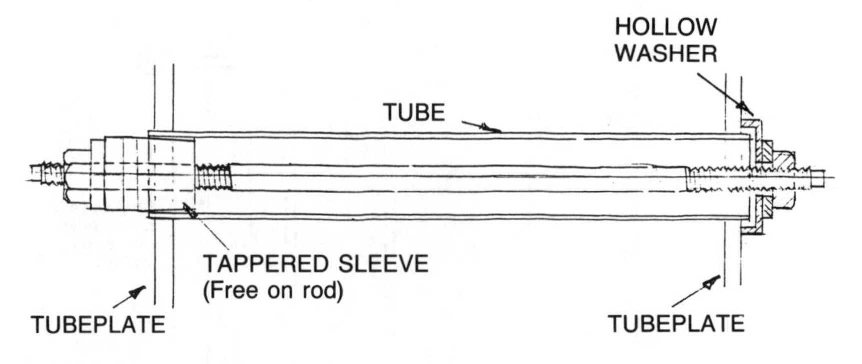

Fig. 12 Tube expander

fitting and expanding the cross-water tubes. Unfortunately, the Allen and Merryweather boilers had no such flat faces and so the cross tubes had to run, parallel to the centreline in plan view across the circular firebox, with the tube ends finish cut at an angle, as shown in plan. Care has to be taken when drilling and reaming the tube holes that they are all lying radially at an angle to the centre of the box.

There are 43 cross-water tubes, 5/16 inch 0/dia. × 22 swg thick, all expanded in place, or as an alternative, silver soldered in place, with Easyflo No. 2. The five vertical water tubes ("J" tubes) each side are positioned radially at 4¼ inch PCD in the crown plate and to make it easier to insert the radiused lower end of each tube into its hole in the plate after inserting the top into the crown plate, the lower end of each

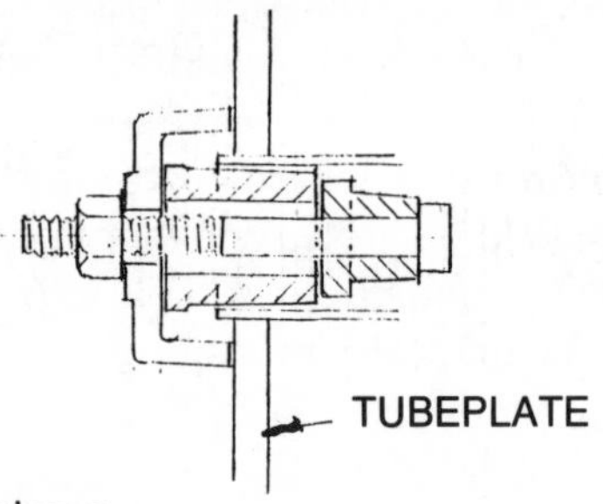

Extracting tappered sleeve with screwed extractor rod and bridge piece.

For use with Tube Extractor.

Fig. 13 Tube extractor

CROWN PLATE
WITH BOSS IF REQUIRED

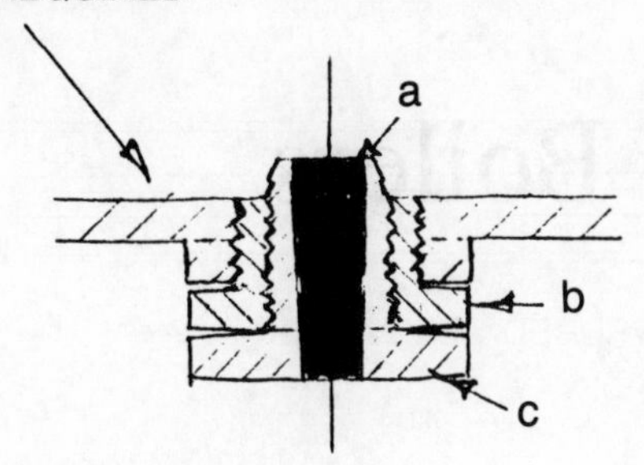

The Boss may be placed on the outside of crown plate if required to give a higher plug position.

a Fusible Metal
b Plug
c Removable Plug

Fig. 14 Fusible plug

tube is bowed slightly. This should make it possible, after annealing each tube, to spring it into the lower hole and ensure a short parallel length for expanding correctly. The vertical tubes also may be either expanded or silver soldered in place.

My standard firehole door can be fitted into the firehole, or a lug can be welded, adjacent to the firehole, on which to mount the firedoor hinges.

Two small angle brackets can be attached to outside of the foundation ring to support the circular cast iron firebars. These should have a space of ¼ inch minimum between the bars.

If any reader decides to use solid drawn steel tube of 6½ inch 0/dia for the boiler shell, all dimensions should be reduced diametrically, in proportion, by ¼ inch and the same water space maintained around the firebox as I have shown. As remarked upon earlier, the top angle ring, welded to the flue, should be drilled and tapped for 24 No. 2 BA steel studs, on 3¾ inch PCD equi-spaced at 0.49 in pitch. The lower angle rings welded to the top and lower sections of the boiler shell should be drilled for 46 × 2 BA hex. steel bolts and nuts, equi-spaced on 7⅞ inch PCD at 0.59 in pitch.

CHAPTER 3

Copper Boilers

The boiler described was designed by me in 1962 and the first two constructed by A. F. Farmer of Birmingham for my own and Colin Tyler's class BB and BB/1 2″ scale ploughing engines. Since then, many more have been built in many different parts of the world, and because of the continuing popularity and interest in these engines the original chapter on construction has been included in this book.

Since designing this boiler I have moved away from crownstays silver soldered to firebox crown and outer wrapper and now always specify a firebox crown reinforced either with the type of stay bent from ⅛ inch copper shown in Fig. 2 or by corrugation or detent in the crownplate itself. It is recommended that this first method is now adopted with the ploughing engine boiler described here, bringing it into line with those designed for the big Superba and class Z7S compounds and the 16 nhp single cylinder engine, all in 2 inch scale also.

The use of this type of crownplate staying makes it even more important that the foundation ring joints should be first class and to ensure a joint of maximum strength the foundation ring should be chamfered all the way round to form a good fillet.

Two further points are the use of 10 swg (.128 inch) thick tube for the barrel and all sheet material used in the boiler, and an increase in the firehole internal diameter to accommodate my liner-fitting firehole door.

The new method of attaching hornplates (Fig. 4), used on my later series of ploughing engines, can equally well be applied to the class BB engine, but as this method eliminates the need for the ⅜ inch × 40 tpi hollow stays, these should be omitted and all stays made 3/16 inch dia × 40 tpi.

Earlier boilers were assembled using Johnson Matthey B.6. alloy and Tenacity 4/A flux for backhead, throat and tubeplates, with Easyflo No. 2 for tubes, stays, etc. However, for some years now, Easyflo No. 2 has

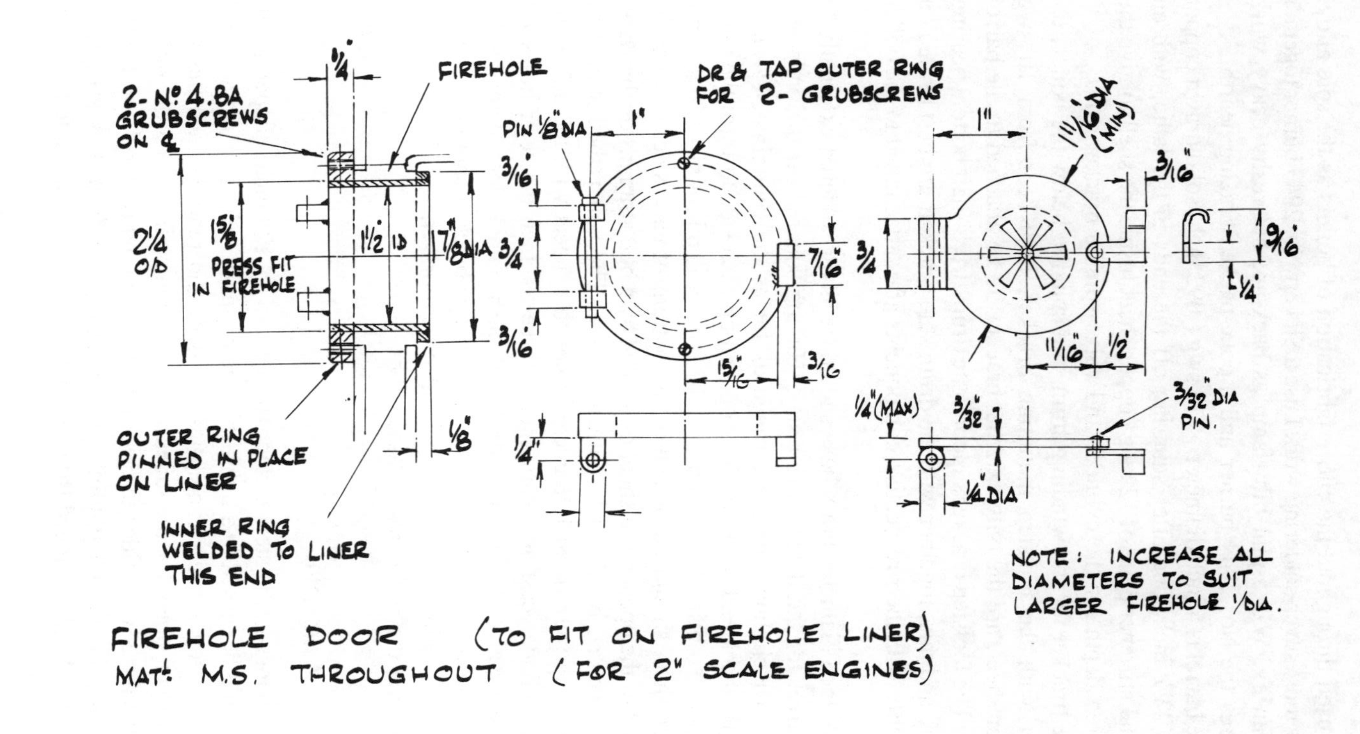

Fig. 15 Standard firehole door

been used throughout the entire fabrication of boilers as the sole alloy.

For most silver-soldering work I use a Sievert No. 2943 burner together with an ex. WD 5-pint blowlamp as back-up, but really heavy work requires the No. 2944 burner and a coke-packed brazing hearth.

The barrel is 5 inch diameter × 13 swg (.092 inch) solid drawn copper tube, 22½ inch long after squaring off the ends in the lathe, with an internal stiffener plate of 12 swg copper sheet inside the shell to give the required depth for the cylinder and valve-spindle bracket studs.

The bracket for the winding-drum stud spindle, below the boiler, together with the footstep mounting bracket, both bent from 10 swg copper sheet, are the only external brackets to be attached to the barrel itself. The footstep bracket supports the drum drive-shaft bottom bearing casting and this, together with the drum stub bracket, are best drilled on assembly with their respective components after attachment to the boiler barrel.

A ½ inch diameter brass boss should be silver soldered externally to the barrel for the drum carrier stay. Having squared off the ends, the boiler barrel should be slit through as far as the centreline, 7¼ inch from the back end, and the sheet opened up each side to form the outer wrapper. To bring the wrapper to the correct depth, a piece of 13 swg copper sheet should be brazed to the bottom of each side, this can be scarfed to the opened-out sheet or jointed by a butt strap inside the water space, secured by 3/32 inch diameter rivets. The boiler shell should be drilled with four 5/16 inch diameter holes to connect with the underside of the cylinder casting steam space, three 7/16 inch diameter holes for the check valve, bushes and top turret bush, and ⅜ inch diameter for the blowdown cock. The stiffener-plate and two brackets should be secured in position with 3/32 inch rivets and either brazed or silver soldered completing the boiler barrel assembly.

Smokebox tubeplate

This is flanged on a steel former from 10 swg (.128 inch) copper sheet, and should be turned to a tight fit in the boiler barrel, after trimming off the rough edges and cleaning-up. Jig drill and ream ½ inch diameter the fourteen holes for the tubes, slightly countersinking on the outside to assist the silver solder to form a fillet on assembly. Drill and tap the four holes for the longitudinal stay blind nipples ⅜ inch × 40 tpi.

The backhead is flanged from 10 swg copper sheet and should be carefully cleaned up to a tight fit in the wrapper plate. Cut the hole for the firebox ring, which must be a tight fit in the plate, and mark off and drill the two ⅜ inch diameter holes for the water gauge bushes.

Drill and tap ⅜ inch × 40 tpi the four holes for the longitudinal stay nipples, and at this stage it is easiest to attach the hinge plate and catch plate for the firehole door, with 2⅛ inch rivets through plate and backhead. An alternative method of attaching these two items is by the use of blind nipples, tapped 4BA and inserted into the backhead.

Throatplate

This is cut from 10 swg copper sheet formed to the inner radius of the boiler barrel, and fitted inside the barrel slit as shown in the drawing, which assists in supporting the plate against the boiler pressure.

Firebox and tubes

The firebox is composed of a single plate, forming the sides and crown, and the two endplates, all in 13 swg copper sheet. The firebox tubeplate should be jig drilled to match up with the smokebox tubeplate, and slightly countersunk after reaming the 14 hole to ½ inch diameter. The backplate is cut for the firehole ring, which again must be a tight fit.

The crownstay consists of two outer girder stays and a centre girder stay, formed as shown on the boiler drawing from 13 swg sheet copper. Before brazing, these should be secured in position by ³⁄₃₂ inch copper rivets, to prevent moving, but great care must be taken to ensure that the rivets hold the flange of the stay tightly against the firebox crown.

The foundation ring is ⅜ inch × ¼ inch copper bar, or ⅜ inch diameter round bar cut and inserted into the water space. The four longitudinal stays are ¼ inch diameter bronze bar, screwed each end 40 tpi to suit the blind nipples, and the 4BA firebox stays are bronze bar – 21 in the backhead, 17 in the throatplate and 42 each side, with six hollow stays each side, screwed externally ⅜ inch × 40 tpi and internally No. 2BA for attachment of the hornplates. All firebox stays are screwed in position and silver soldered, as described later. The bronze bar under the trade name of *Hidurel* 5 or 6 gives the maximum strength–stress in tension, of 42–52 and 28–34 tons per square inch and is suitable for both longitudinal and firebox stays.

On no account should either steel or brass rod be used for any stays.

Fig. 16 Fowler boiler

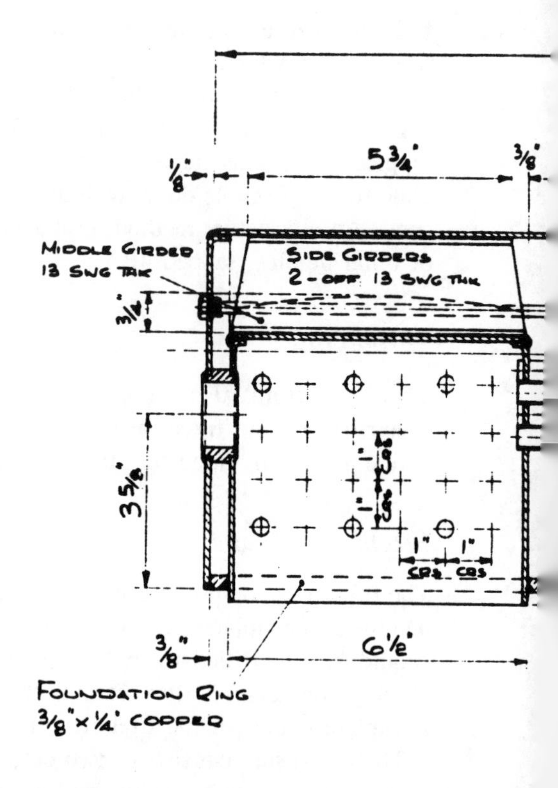

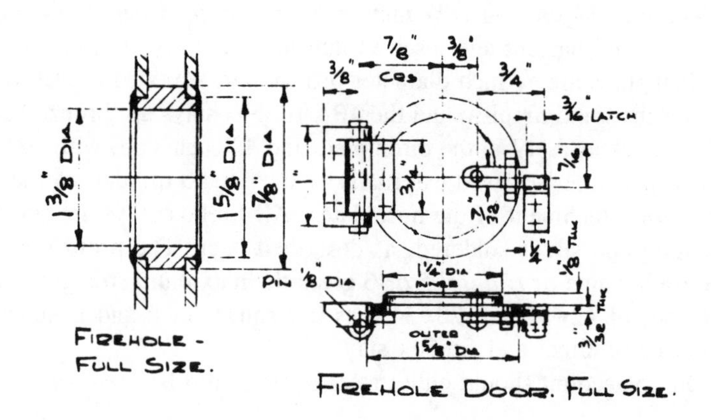

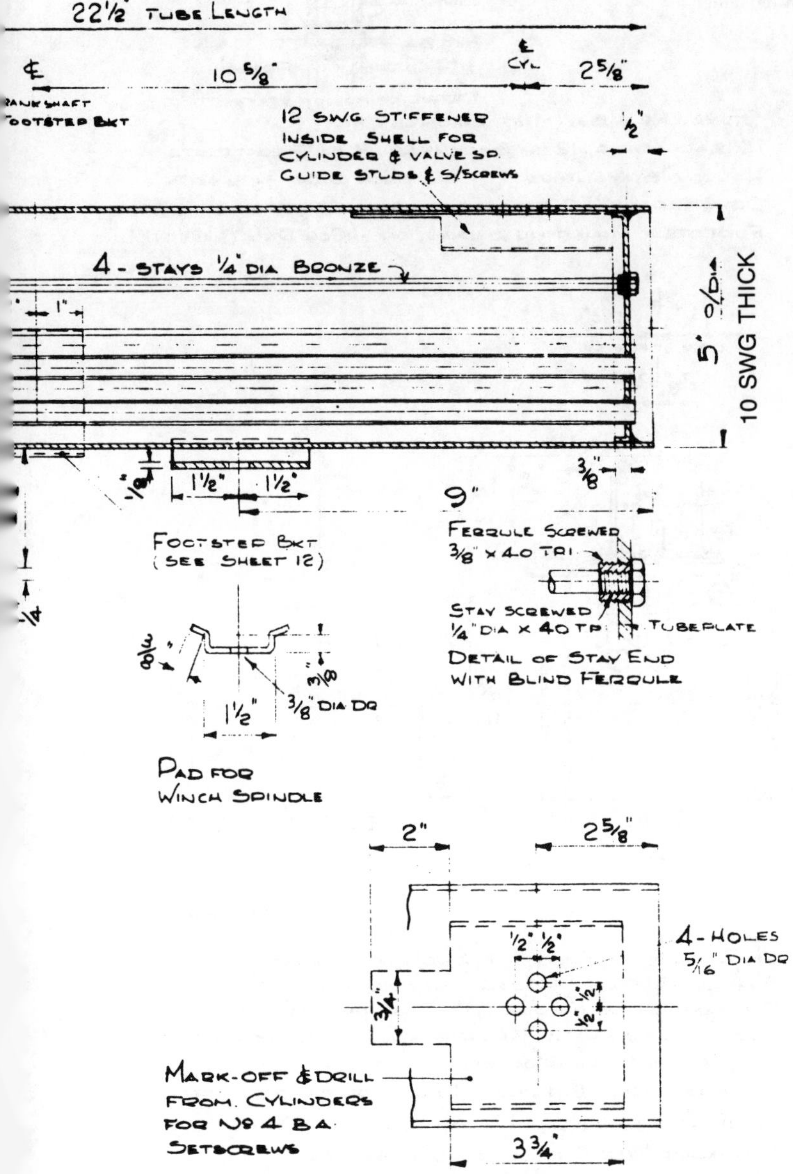

22½" TUBE LENGTH
℄ CRANKSHAFT FOOTSTEP BKT
10⅝"
℄ CYL
2⅝"
12 SWG STIFFENER INSIDE SHELL FOR CYLINDER & VALVE SP. GUIDE STUDS & S/SCREWS
½"
4 - STAYS ¼" DIA BRONZE
5" O/DIA
10 SWG THICK
1"
1½"
1½"
9"
⅜"
FOOTSTEP BKT (SEE SHEET 12)
FERRULE SCREWED ⅜" × 40 TPI
STAY SCREWED ¼" DIA × 40 TPI
TUBEPLATE
DETAIL OF STAY END WITH BLIND FERRULE
⅜"
⅜"
1½"
⅜" DIA DR
PAD FOR WINCH SPINDLE
2"
2⅝"
4 - HOLES 5/16" DIA DR
½" ½"
3¼"
MARK-OFF & DRILL FROM CYLINDERS FOR Nº 4 BA SETSCREWS
3¾"

Fig. 17 Fowler boiler

SILVER SOLDER
Nº 2 B.A.
3/8" x 40 T.P.I.
FIREBOX
DETAIL OF HOLLOW STAYS

STAYS : Nº 4 B.A. MATL BRONZE.
18 EACH SIDE. 12 IN BACKHEAD. 8 IN THROATPLATE
HOLLOW STAYS THUS ⊕ 6 - EACH SIDE 3/8" x 40 T.P.I.
DR. & TAP Nº 2 B.A FOR SETSCREWS SECURING HORNPLATE.
FOOTSTEP MOUNTING SHOWN ON WINCH DRG (SHEET 12)

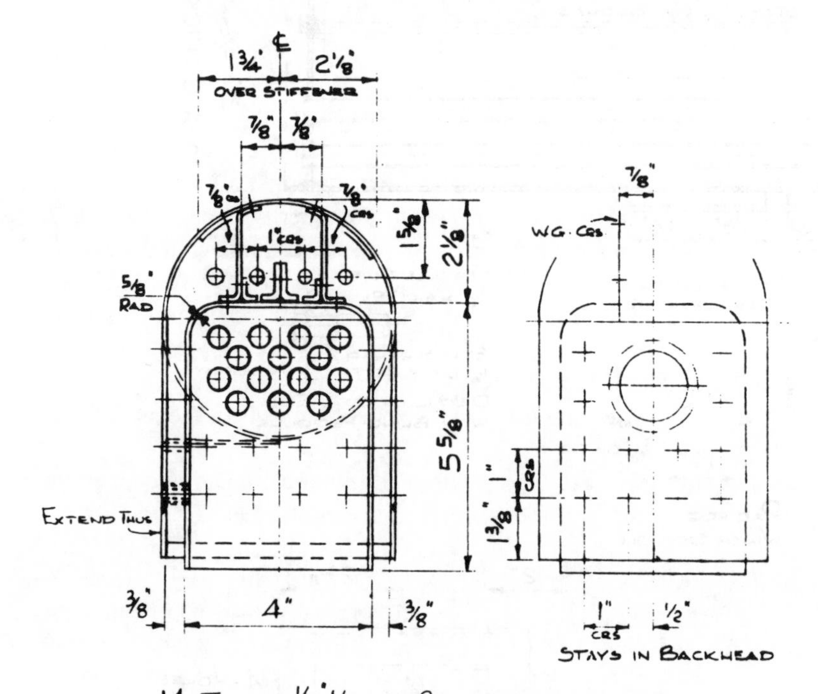

14 TUBES 1/2" O/DIA x 20 SWG THK COPPER
SHELL 13 SWG THK S.D COPPER TUBE 5" O/D.
FIREBOX SIDES & CROWN 13 SWG THK COPPER
FRONT TUBE PLATE, THROATPLATE, AND BACKHEAD
ALL 10 SWG THK COPPER
CONSTRUCTION : BRAZED. TUBES SILVER-SOLDERED
WORKING PRESSURE 75 LBS /□"
HYDRAULIC TEST PRESSURE 150 LBS /□"

FOWLER COMPOUND PLOUGHING ENGINE 2" SCALE
16 NHP CLASS B.B. & B.B.I.
BOILER AND FITTINGS

The fourteen ½ inch diameter tubes are 20 swg copper cut to a length of 15⅜ inches.

Notes on construction

The boiler barrel should now be complete, with wrapper extensions and throatplate brazed in position, and awaiting the fitting of the firebox and remaining items.

Assemble and braze up the firebox, with firehole ring inserted and slightly peened over where it fits through the plate, and crownstays in position. Insert the tubes into the tubeplate, and line up with the smokebox tubeplate, and silver solder in position. Next, insert the firebox into the boiler, and secure firebox tubeplate to throatplate with three 3/32 inch copper rivets, through both plates and the front length of foundation ring.

Rivet the top flanges of crownstays to the outer wrapper plate. Insert the smokebox tubeplate into the barrel – this should line up if previously checked with the tubes before assembly – expand the tubes into their reamed holes in the tubeplate, and silver solder the flange of the tubeplate or boiler barrel, and, next, the tubes into the tubeplate.

Now silver solder the crownstays to the outer wrapper. The firebox and tube assembly is supported in the barrel by means of the crownstays and front tubeplate.

The backhead is next inserted into the wrapper plate and over the firehole ring spigot, and the next job is to ensure that the wrapper fits closely to the backhead flange and that the firehole ring spigot is tightly peened or riveted over the plate.

Fit the remaining lengths of the foundation ring, ensuring that they are a good fit all round, and secure in place with three 3/32 inch copper rivets each side. Silver solder the backhead into the wrapper, the foundation rings, the various bushes, and the longitudinal stays.

The side and end stays may be tackled next. They must be a reasonably tight fit when screwed home, and may be riveted over both ends and silver soldered or inserted and silver soldered over head and nut. The hollow stays are screwed home tightly, and then silver soldered – cut off the spare length on assembly. Before fitting the hornplates, the protruding stay heads will require machining or grinding to uniform length.

A point worth mentioning is that on the full-size engine, the top water gauge connection is led through the centre of the circular bracket, on top of the backhead. It is better to drill and tap all bushes before silver soldering into the boiler, merely running a clearing tap through them after assembly to correct any slight distortion of the thread due to heat.

The ½ inch diameter boss just forward of the winding drum stub bracket should also be tapped No. 2BA before attachment to boiler, for the drum guard stay.

CHAPTER 4

Testing and care of Boilers

Testing

As mentioned in the previous chapter, a welded steel vessel may be pre-tested after fabrication by swilling a small quantity of paraffin along all the internal welded seams; any minute weeps or cracks will quickly become apparent as the thin oil spreads through to the outer surface.

It may save much time and extra work if sub-assemblies of a welded boiler are separately tested thus before final assembly, but all traces of oil should be washed away with hot water and, if necessary, a mild detergent afterwards.

My own hydraulic test pump is built up from a 1¾ inch bore by 4 inch stroke cylinder, with integral tank holding one gallon of water, a 6 inch dia dual reading gauge and ¼ BSP water connections and valves.

It is sometimes necessary to apply gradual pressure by only a few pounds at a time when approaching the test figure, particularly with a small boiler, and as this pump has a healthy delivery at full stroke, a small secondary cylinder is mounted alongside the main one. A bypass valve diverts delivery from one to the other as required.

A reliable, accurate and easily read pressure gauge is a vital necessity, and the cheaper variety should be avoided like the plague. A minimum face diameter of 5 inches, reading up to at least 300 psi, should be used, with the gauge firmly mounted on a board or bracket above the pump where it is protected from vibration. In addition, a gauge can be mounted directly on the boiler if required. Prior to putting a steel boiler into service, and when it is absolutely dry inside, a small quantity of powdered graphite should be fed through any available opening. As a rough guide a vertical boiler 8½ inches dia. × 12 inches high required just over ¼ pound of graphite to coat the internal surface. Shake or revolve to spread the powder evenly inside, as it will help to prevent lime scale and sediment forming and should be repeated annually after the hydraulic test, or after a full wash-out.

Anti-corrosive measures

Satisfactory feed water treatment for boilers of road and agricultural engines, which are liable to pick up water from widely differing areas, has always been a problem; treatment suitable in one locality is quite likely to be incorrect and possibly harmful in another.

After seeking advice from the experts, I decided against using any chemical additives in my steel boilers, and to follow instead the suggestions given to me.

1. Most corrosion arises from oxygen pitting caused by dissolved oxygen in the boiler and feed water. Feed water should be pumped into the boiler above the water level, as oxygen is then released into the steam space and dissipated with the steam passing up to the cylinders, without coming into contact with platework below the water level. Top feed, then, is preferable.
2. Hot water passed into the boiler from an injector is preferable to cold water from a pump, as oxygen is more readily separated from water when it is hot. Thus an exhaust-steam feed water heater is a help, if only a pump is fitted.
3. Do not use water that is acid. Rain water, though it is usually assumed to be the purest, can be slightly acid due to the chemicals picked up by rain as it falls. This is probably likely to be worse in certain areas.
4. Avoid too frequent washing-out. Water which already contains some corroded iron (red iron oxide), is slower to dissolve more.

Laying-up a boiler

As protection against boiler corrosion during the winter is a subject on which varying views are held, I will merely list the measures I have used on my own boilers for the past twenty-five years or so to minimise oxygen corrosion inside and out.

1. After the last steaming before winter, blow the boiler down when still really hot to remove all traces of damp inside.

2. Remove safety valve or filling plug and suspend inside a small perforated bag of silica-gel. Replace valve or filling plug and leave the boiler completely closed up.
3. Clean inside the firebox and smokebox thoroughly with a stiff wire brush and paint with anti-corrosive aluminium paint or mineral oil, brushing well into all corners and around tubes. It is not advisable to use vegetable oils as traces remaining can, when heated, turn acid and attack plates and welds.
4. When preparing the boiler for work the following spring, avoid filling with water too soon before first steaming as the rate of corrosion is slower once some scale has formed and oxygen boiled out of the water.
5. In between steamings the boiler is best kept absolutely full, with no air spaces at all. draining out to bring down to working level before lighting up.

As long ago as 1969 *Model Engineer* published, in issue number 3379, an abbreviated testing procedure for model boilers which was intended as a general guide for constructors, users and model engineering club officials.

Eight clauses of this procedure have been reproduced in their original form, but clauses 3, 7, 8, and 12 have been re-written to bring them into line with testing procedures used by me in my professional capacity.

Model boiler testing procedure

(reproduced by courtesy of the Editor, *Model Engineer*)

1. It is recommended that all model boiler testing shall be carried out by two competent members of the model engineering society concerned.
2. The boiler inspectors should preferably, but not necessarily, be qualified engineers, but they should at least be men with both theoretical knowledge and practical experience of steam models.
3. The boiler should be stripped of lagging and all insulating material and cladding to expose joint faces and stay heads; if necessary the smokebox and hornplates may be removed.
4. An hydraulic test with cold water should then be carried out at twice the normal working pressure. The test pressure should be applied at least three times, the pressure being allowed to return to normal

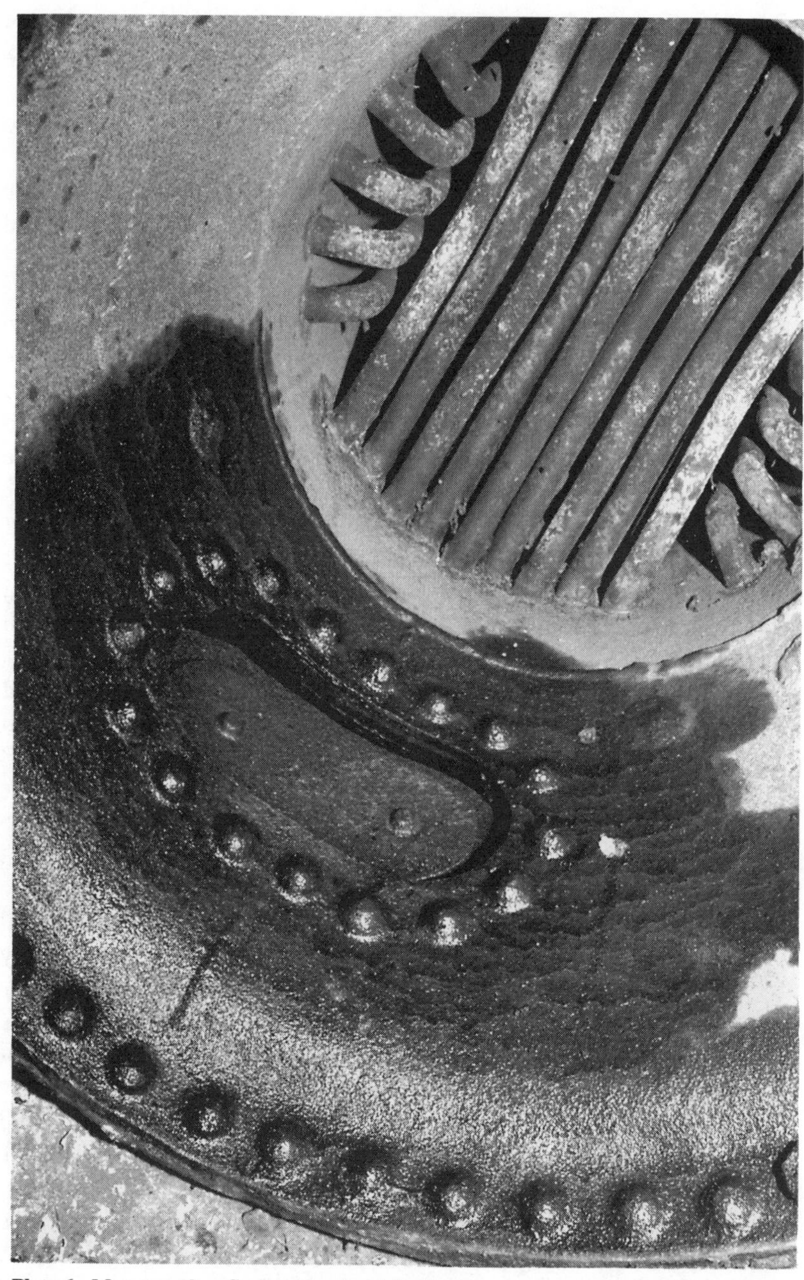

Plate 6 Merryweather firebox showing tube layout

Plate 7 Airborne steam generator

between each test. After the hydraulic test, the boiler should be thoroughly examined for collapse or distortion. While appreciable distortion will condemn the boiler, slight leakage need not do so, as this can nearly always be corrected, after which the boiler can be again tested.

5. Safety valves must be of sufficiently large area to prevent the boiler pressure rising more than 10% above the working pressure, however hard the boiler is worked.
6. The boiler inspectors should examine all the fittings on the boiler during the second and subsequent tests.
7. Steel boilers to be tested every twelve months, copper boilers every two years.
8. After the boiler tests have been satisfactorily completed the boiler inspectors should issue a suitable certificate, signed by them, and dated, and marked "valid for ... months only". A register of boiler tests to be maintained, each boiler and its certificate to be duly numbered.
9. Although not essential, a steam test of the boiler, with all fittings in place, at a pressure 10% higher than working pressure, is recommended.
10. If the boiler has been built to other than a recognised published design, the boiler inspectors should examine the drawings and details of construction, before accepting the boiler for test.
11. All the above clauses apply to copper boilers with brazed or silver-soldered joints as well as steel-welded boilers.
12. Steel and copper boilers with a working pressure of up to 100 psi to be tested to **twice** the working pressure; above this working pressure discretion should be used as it is bad practice to deliberately aim for an unnecessarily high test pressure, with the attendant danger of plate malformation and internal damage to the boiler.

Many users of steam plant insist on test figures of only a small percentage above working pressure if this is above 125 to 180 psi.

In the case of traction engines constructed with separate horn-plates i.e. not upward extensions of the outer wrapper plates, these should be removed before testing, together with any false throat or backhead plates.

Where possible, the steel boiler plates should be measured for signs of wasting.

Some boiler data

Cylindrical boiler shells (steel)

Maximum permissible working pressure formulae:-

$$P = \frac{2fE\ (T - .06)}{d + (T - .06)}$$

Where P = max. permissible working pressure
d = internal diameter of shell in inches
T = thickness of plate in inches
E = joint factor (1.00 for an unpierced seamless shell)
f = maximum permissible working stress in lbs. per square inch.

Joint factors (E)		*Max. permissible stress (f)*
Single butt weld	= .35 up to 650° F	= 14,500 lb/sq. inch
Single butt weld with backing strip	= .55 up to 750° F	= 12,750 lb/sq. inch
Double sided butt weld	= .60 up to 900° F	= 6,300 lb/sq. inch
Stress relief by heat	= .10 (Steel at 26/30 tons/sq. inch)	

Cylindrical boiler shells (copper)

In the design of copper boilers of all sizes, I have always used the following formulae, attributed to the late Henry Greenly:

$$\text{Working Pressure (WP)} = \frac{S \times P \times 2 \times R \times C \times T}{D \times F}$$

where S = Ultimate strength in lb/sq. inch (for copper 16,000 minimum)
P = Plate thickness
F = Factor of safety of from 6 to 10
R = Riveting allowance .5 for single
.75 for double
.8 for welded, brazed or silver soldered
D = Diameter of barrel in inches
C = Corrosion allowance (for steel only) 1/4 thick and below = .5 to .8
T = Temperature allowance Copper at 400° F = .7
Copper at 212° F = .87

For a given diameter of boiler, the plate thickness would be:

$$P = \frac{D \times WP \times F}{S \times 2 \times R \times C \times T}$$

Firebox stays should be twice the thickness of the inner firebox wrapper plate, in diameter, not forgetting that the effective diameter is that of the core, and not over the threads; the pitch of firebox stays should not exceed:

$$\frac{t \times 800}{WP}$$

where t = thickness of firebox inner plate in inches and WP = working pressure in lb/sq. inch.

Brazing requires well fitting flanged joints with no gaps, whereas bronze-welding requires fairly loosely fitting joints with well-rounded flanges.

Oxy-acetylene flame should *never* be used for brazing with any alloy containing copper-zinc, owing to the zinc "boiling out" at high temperatures, leaving the joints porous and weak.

Alloy	*Makers*	*Melting Point °C*	*Tensile Strength Flux in tons/sq. inch*
Sifbronze No.1	Suffolk Iron Foundries Ltd.	850°	28. Sifbronze
Brazotectic	B.O.C.	875°	29.1 Brazotectic
:Cuprotectic	B.O.C.	705°	10.7 None
Silbralloy		638°–694°	35. None
*Argoflo	Johnson	605°–651°	32. Easy-flo
*Easy-flo	Matthey &	620°	30. Easy-flo
Sil-fos	Co. Ltd.	625°–780°	45. None
+B6		790°–830°	28. Tenacity 4/A

Characteristics:

:Low elongation figure, and apt to fail through brittleness under varying temperatures.

*Recommended for screwed stays etc., and tube ends.

+Recommended for main construction – backhead, throat, and tubeplates for smaller boilers, Easy-flo may be used throughout.

Flat-stayed surfaces

The thickness of the front tubeplate, backhead, and other flat surface plates will always be more than that of the boiler shell, as a glance at some of the illustrations will show.

As a rough guide for those that do not wish to become too involved in calculations, all flat-stayed surfaces should be at least one third thicker than the boiler shell material, unless for availability reasons some very heavy material has been used for the shell; it will be hardly necessary to emphasise, though, that all flat surfaces must be stayed, unless, as in the case of a vertical boiler, the firebox tube itself acts as a robust staying element.

Stays are, of course, not necessary in the case of the circular firebox type of boiler, except as longitudinal stays between the boiler ends.

On larger boilers, front tubeplate and backhead are sometimes strengthened by the addition of gusset plates between the inside surface of the flat plate and the inner radius of the boiler shell, but this appears to be seldom used or necessary on smaller boilers. A circular boss can be a stiffener and flat surfaces under pressure may be stiffened by formed indentations or even slight inward doming, but these methods of obtaining extra strength are usually difficult to incorporate in the design of, say, a locomotive boiler in small scale, except in the case of firebox crown plates. The total load on, for example, the end of a five-inch diameter plain cylindrical boiler at a working pressure of one hundred pounds per square inch would be just over 1,900 pounds, and this force, acting both on the plate itself and any joint between the barrel and the plate, illustrates sufficiently clearly the necessity to stay any such plate. In fact, tubes themselves help to support the tubeplates, the load being shared with two or three longitudinal stays.

Tube lengths and diameters

Boiler tube length, in the case of locomotive type boilers, should be governed wherever possible by the formula:

$$\text{Tube length} = \text{Internal dia}^2 \times 50 \text{ to } 70$$

The external diameter of tubes in full-size boiler practice range from 2 inches to 3½ inches, and those for model work vary considerably – I find that ½ inch internal diameter is the minimum for a boiler such as the 2-inch-scale Fowler ploughing engine, this size permitting an adequate flow of hot gases while not sifting up with ash too quickly. Too large a tube diameter permits the hot gases to flow through to the smokebox too readily, a nearly red-hot smokebox not being conductive to efficient boiler management.

In practice, the Fowler boiler mentioned above will continue to steam easily and freely with the bottom row of tubes completely blocked by ash deposits, as onlookers at more than one rally will testify, and I have views as to the value of comparatively few tubes of ½ diameter compared to a larger number of, say, 5/16 to 3/8 inch diameter.

Boiler materials

The steel plate used in manufacturing the boilers described in these pages is medium carbon (0.3% to 0.5%) steel to British Standard Specification 968 or 4360.

Copper tubes are normally made in de-oxidised copper, either non-arsenical or arsenical. De-oxidised copper has been freed from dissolved oxygen by the addition of a de-oxidising agent, usually phosphorous.

The manufacturers recommend that copper treated in this way is used when high melting point brazing material is to be used in fabrication of the boiler or vessel.

The presence of a de-oxidising agent lowers the electrical conductivity of the copper. Arsenical copper has had an addition of arsenic usually around .4 per cent which has been made for the purpose of increasing its strength and toughness at high temperatures.

Fusible plug

In the case of full-size engines the practice of fitting a fusible plug to protect the firebox crown from damage resulting from low water level was almost universal.

With the increase in the use of larger steel boilers this practice could well be extended to include boilers of say one-third full-size upwards.

Plugs were supplied to boilermakers and engine builders by the manufacturers of boiler fittings and mountings, the most commonly used type being Saxon's fusible plug which utilized easily replaceable lead discs, or a type with a lead-filled conical nose, detachable for refilling with molten lead (see Fig. 14).

Provision of a fusible plug necessitated the inclusion of a boss in the centre of the firebox crown, drilled and tapped to suit the plug thread, which was usually slightly tapered. This will be an easy modification to include in the design of a modern all-welded firebox, using either a ¼ inch BSP thread or a suitable ME fine thread.

Lighting-up and raising steam

Ideally, boilers should be allowed to build up from cold to working pressure as gradually as possible.

A small electrically driven fan unit designed to rest on the chimney top is an indispensable aid to lighting-up, especially on the smaller scale engines, but care should be taken to see that the induced draught is not too fierce, resulting in a boiler being "forced".

Blower units built up from ex-car heaters are likely to be a little on the fierce side. It is not necessary to use an electric blower with the Suffolk dredging tractor steel boiler, for instance, as an extension about three feet in length fitted to the chimney soon pulls the fire up sufficiently to give a few pounds pressure through the blower valve. The big firebox enables a good wood base to be laid – the old engineman's favourite, ash, gets a fire going quicker than anything even with inferior coal; and keeps the inside of the box cleaner than using paraffin soaked box-wood or firelighters.

Avoid using lighting-up wood dripping with paraffin, or throwing paraffin onto a slow fire. If "natural" wood such as ash is not available, soak small pieces in paraffin for a day and then allow the surplus to drain off before using them. This avoids sooting up the tubes unnecessarily.

In small fireboxes such as those of the 2 inch scale Aveling roller and Ransomes tractor, a few lumps of charcoal mixed with the first layer of coal will help to get a good fire going with the minimum use of the electric blower.

Lady Windsor small steam nuts light easily and make a very hot fire, but this coal is expensive and burns rather quickly. In large fireboxes such as the Suffolk and Fowler Superba I use a handful of these nuts to start the fire and then run entirely on a brand of domestic boiler fuel known as dry steam semi-anthracite, in lumps about one inch square.

Some of the steam coal supplied at rallies for full-size engines breaks down into manageable small lumps and mixes well with the semi-anthracite, adding a trace of smoke to the smokeless fuel. Generally, household coal should be avoided, particularly the variety containing thinly disguised lumps of slate.

Firebars should be spaced rather wider than usual if anthracite is burned – the Suffolk, Durham and N. Yorkshire traction and all the Fowlers have 5/16 inch spaces between bars – and despite its reputation I have not found that this fuel is unduly hard on either steel or cast iron firebars.

Tubes should be swept and firebox and smokebox cleaned out after a prolonged steaming, and it helps to keep corrosion at bay if both the ashpan and smokebox are wiped round with an oily rag after cleaning out the ash – but as mentioned earlier, don't use a vegetable oil.

Smokebox doors are sometimes difficult to keep one hundred per cent airtight, and the thick black mastic used in sealing cracks in roots or walls is very useful for sealing around edges or in the areas of hinges. An internal baffle plate, fitted inside the smokebox door, helps to keep the door from becoming too hot – particularly in a short smokebox. Many of the replacement smokeboxes fitted to engines by John Allen and Son (originally the Oxford Steam Ploughing Co.) had a deflector baffle fitted above the tubes as well as one inside the door.

CHAPTER 5

Safety valves

A safety valve particularly favoured by traction and portable engine makers for many years, was the Salter type in which the valve body was held down upon its seat by a lever pivoted in a bracket at one end, and pulled down at the other end by a spring encased in a brass sleeve, which also contained a pointer to indicate the spring loading. An adjusting screw and locknut allowed adjustment to the spring to be made, and sometimes this was tampered with by a driver seeking greater boiler pressure. These Salter valves were usually encased in a handsome bell-mouthed housing.

Gradually the spring balance valve gave way to other types on locomotive, traction and portable engines, although for a number of years Fowlers continued to use the spring balance type occasionally on some ploughing engines built for overseas. Very widely used was the Ramsbottom spring-loaded double safety valve which originated from the Crewe works of the old LNWR, and land boilers such as the Lancashire or Cornish type were fitted with deadweight valves in which a series of heavy weights was mounted directly over the valve itself, usually housed in a plain casing. Another simple form of safety valve had a lever with adjustable weight suspended from the end, resting across the top of the valve and retaining it upon the seating – this latter example once very popular for stationary vertical boilers was unsuitable for use on agricultural and other self-moving engines where spring-loaded valves became the rule.

An advantage of the spring-loaded valve over the deadweight type is that the heavy weights needed for high pressures are dispensed with. Also it is not so liable to stick, and access is easily gained to the moving parts.

A disadvantage is that as the valve lifts from its seating, the pressure exerted by the spring increases, and if the valve is to remain open, the boiler pressure must rise too. To overcome this problem, spring loaded valves of the ''pop'' type came into use, in which an extended bent lip or annular recess overhangs the seating, providing an increased area upon which the steam is effective until boiler pressure is reduced, when the pop valve closes smartly back on to its remaining seating without ''dribbling''. Legislation covering the design and construction of safety

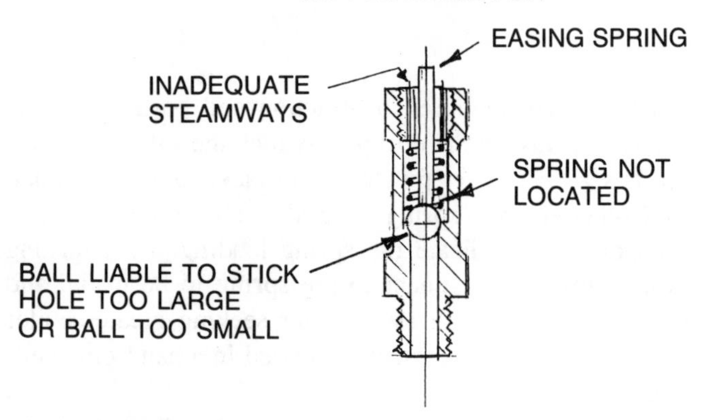

A badly designed simple safety valve

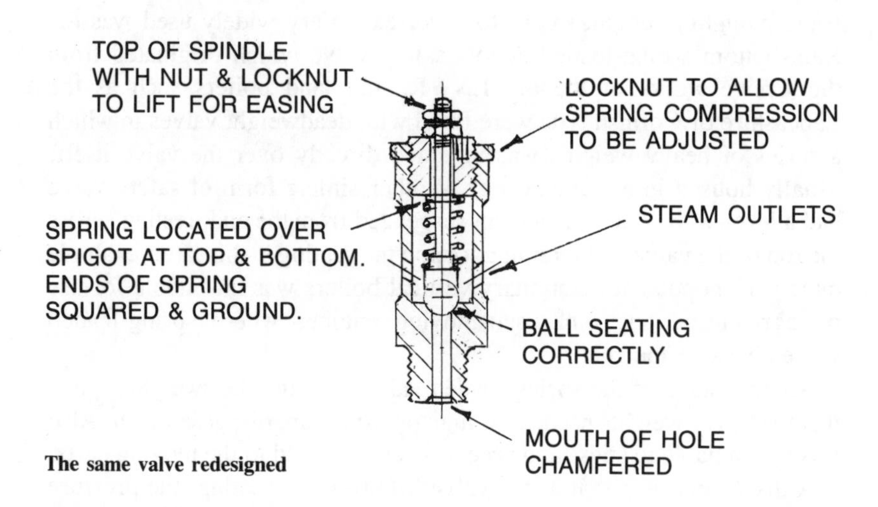

The same valve redesigned

Fig. 18 Safety valve

and relief valves was introduced many years ago, and the Board of Trade, which incidentally had to investigate every case of pressure vessel explosion, laid down strict rules covering safety valve requirements, as did Lloyds Register in the case of marine boilers. Certain of these requirements apply equally to valves used in model engineering.

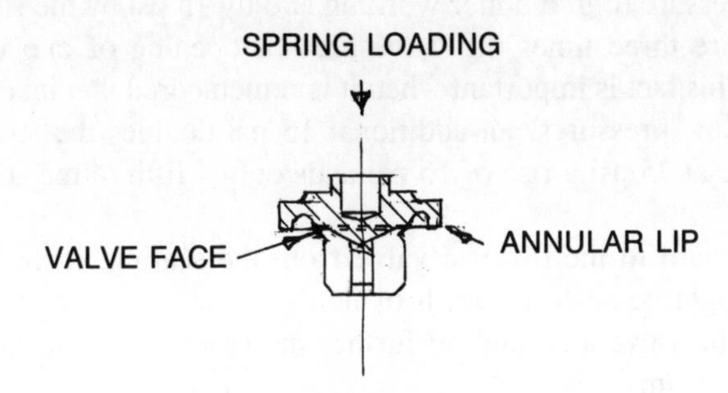

Section through typical "pop" valve

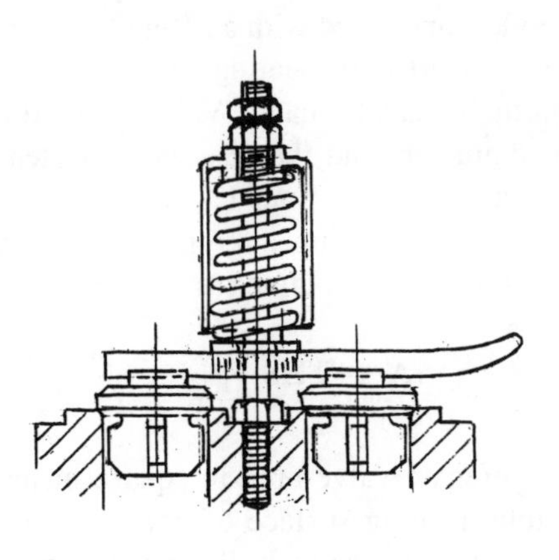

Typical Fowler safety valve shown in part section

Fig. 19 Safety valve

The most important facts to bear in mind when designing safety valves for a boiler of any size are that the steam production of a boiler is sensibly the same at all pressures, and since the density of steam increases very nearly as the pressure, and since steam escapes from an outlet at practically constant volume, safety valves, like boiler stop valves, may be reduced in sizes as pressure increases; consequently a boiler working at a low pressure will require larger safety valve areas than one working at high pressure (e.g. a boiler working at only 15 psi by the steam guage will require three times the valve area of opening of one working at 75 psi). This fact is important when it is remembered that in a boiler safe for only low pressures, an additional 15 psi doubles the strain, whilst in a boiler at 75 psi a rise of 15 psi adds only a fifth to the stress on the plates.

With regard to the lift of a valve from the seating, if the valve lifts only a height equal to a fourth of its diameter, the area of opening is equal to the valve area and no further increase in lift will give greater output of steam.

In the case of a well-designed deadweight valve opening at a pressure of 60 psi, a pressure of 70 psi should keep it open enough for the escape of all steam produced.

Board of Trade rules laid down that safety valves must be mounted directly on the boiler, and fitted with a lifting lever or other arrangement for easing the valves off their seating, which must also be capable of being rotated on their seats by hand. Valves should have a lift equal to one-fourth their diameter, and the passages for steam to and from the valve seats an area of not less than 1.1 times the aggregate valve area in square inches. The minimum aggregate valve area of safety valve should not be less than that obtained from the following formula:-

$$A = THS \times \frac{K}{p + 15}$$

where A = aggregate valve area in square inches
THS = total heating surface of boiler in square feet
p = working pressure in lb. per square inch
K = a constant ranging from 1.1 to 1.5 according to type of boiler and means of firing.

After installation, valves must be given an ''accumulation test'' under full steam and full firing conditions, with feed water shut off and stop valve or regulator fully closed for 15 minutes (7 minutes for water tube boilers). The pressure accumulation must not exceed 10 per cent of the figure at which the valve is set to blow off.

Harking back to the formula for finding the area of safety valves, and applying it to a model locomotive type boiler, for example of 5 inch diameter with ten ½ inch diameter tubes and a firebox totalling 288 square inches of heating surface, we find that for a working pressure of 100 psi the required safety valve area works out at .026 sq inch, giving a valve diameter of 3/16 inch. To allow for use at lower pressures this could well be increased to as much as 5/16 inch or 3/8 inch diameter, and care should be taken to ensure that the steam can get out of the valve housing or casing freely and without restriction.

A valve constructed as shown in Fig. 18 is badly designed as it restricts steam outlet and furthermore cannot be eased off the seating for testing or blowing through. if a ball is used for the actual valve, it should be of stainless steel and should not be capable of jamming or sticking in the seating. The spring should be located at the top, and preferably should not rest direct on the ball at the lower end. Fig. 18 shows a better design. There are a number of suppliers of small rust-proofed steel springs, with ends accurately squared and ground, suitable for model work, and it is far easier to select one by trial and error than to attempt to calculate the spring requirements for small-scale work. My personal preference is for ball valves rather than wing valves for 2 inch scale and under, with all steamways on the generous side and valve housings arranged to vent upwards rather than through outlets in the side – if only because this looks far tidier and makes less surrounding mess!

Lastly, if a boiler is likely to be out of use for a considerable time it is not a bad idea to remove the springs, or at least relieve tension or compression on them, to avoid fatigue. With a valve as shown in Fig. 18 it is an easy matter to relieve the springs of all compression load by simply screwing back the top locknut, and turning the screwed spindle-housing until pressure on the spring is removed, adjusting down again to the correct pressure when the engine is put into service once more.

Plate 8 The two inch scale Aveling and Porter class AD copper boiler

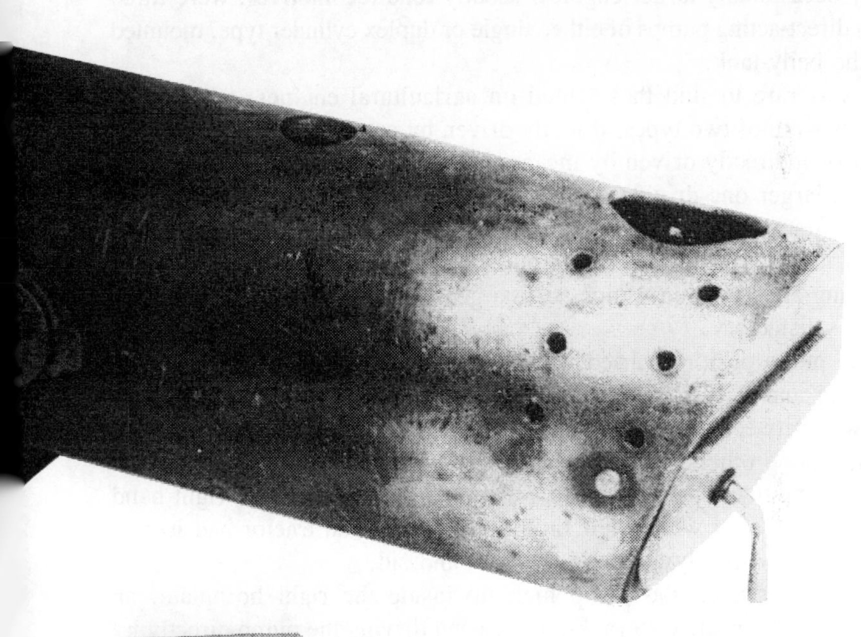

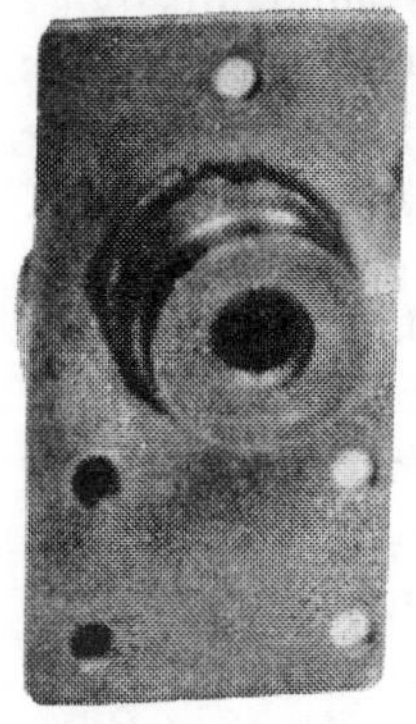

CHAPTER 6

Boiler feed pumps and engine notes

Very occasionally larger engines, usually road locomotives, were fitted with direct-acting pumps of either single or duplex cylinder type, mounted on the belly-tank.

It was rare to find these fitted on agricultural engines, where feed pumps were of two types, directly driven by an eccentric on the crankshaft or indirectly driven by means of a small crankshaft gear meshing with a larger one driving the pump eccentric.

The first was quite satisfactory if the crankshaft speed was not too high, but if it was, a reduction in the pump speed was essential for efficient working and in 2 and 3 inch scale a geared down pump is usually far more satisfactory.

The pump position varied widely from maker to maker; some, like Aveling and Porter, preferred to mount the pump on the boiler side midway between crank and cylinder positions, while Marshalls and Allchin, among others, preferred a footplate position.

Ransome traction engines had a direct driven pump on the right-hand side of the boiler while their little 4 nhp compound tractor had it gear driven below the flywheel on the left-hand side.

Fowler mounted the pump high up inside the right hornplate, an eccentric just inside the crankshaft bearing driving the pump directly by means of a rod bent to clear the second motion-shaft. On certain classes of ploughing engine a pump eccentric was dispensed with altogether, one valve eccentric serving to drive the pump from the back half to the strap.

The Durham and North Yorkshire 6 nhp traction engine was fitted with a gear-driven pump mounted low down on the left hornplate and it is likely that due to the close proximity of the firebox the driver would have had to resort to the old trick of tipping an occasional bucket of cold water over the pump to cool it off. The 2 inch scale Durham follows the same layout but the pump is carried on a separate mounting plate which helps to keep it a little farther away from the hot firebox; with a bore of ⅜ inch and ½ inch stroke this pump keeps the 5 inch diameter boiler well supplied under working conditions.

In selecting or designing a pump the following tips may be useful:

1. Never use a boiler clack-valve with a larger valve area than that of the pump delivery valve as the effect of boiler pressure on the larger area will absorb more engine power in driving the pump.
2. Keep both the suction and the delivery valves as close as reasonably possible to the end of the pump cylinder.
3. To increase pump delivery, lengthen the stroke rather than increase the pump bore. The variable-stroke pump fitted on the Stuart Turner wartime steam charging set was a good example of this.
4. If 0 rings are used instead of graphited packing on plunger and pump gland, stick to the maker's recommended groove dimensions carefully.
5. When mounting the pump, secure to hornplate or boiler facing with setscrews through the pump body, but take all the considerable load of the pump thrust on one or preferably two steel dowels, not less than 3/16 inch diameter, let into pump body and mounting.
6. Use corrosion-resistant steel balls, or even better, phosphor-bronze ones for suction and delivery valves, and make sure that balls are free to move off and back on to their seats.

I use two standardised pump designs for all my engines, merely altering the angle of the pump cylinder relative to the valve box to suit 2 inch scale engine layouts, and using the larger horizontal pump fitted on the Suffolk dredging tractor for larger engines. (Figs 20 and 21)

Injectors

If an injector is fitted it should be positioned low down on the outside of the tender or hornplate, about level with the water tank bottom and far enough from the firebox and ashpan to keep it as cool as possible. A medium size, delivering about 1¼ pints per minute is recommended for a 2 inch scale engine. Steam should be taken via a globe valve from a turret on the boiler top just forward of the backhead. The turret should also have a connection for the pressure gauge, and an additional connection screwed ¼ or 5/16 × 32 tpi is always useful if steam is required for any other purpose. This can be blanked off with a screwed plug until it is required.

Water lifter

After experimenting with various internal shapes and sizes, a reliable water lifter was developed for the 2 inch scale Durham and North Yorkshire engine and the Fowler ploughing engines in the same scale. Using a discarded standard ¼ BSP threaded Maxam air fitting, modified as shown in (Fig. 22) resulted in a very efficient water lifter, very slightly over-scale but lifting water at little more than 20 psi pressure to the accompaniment of just the right noise. Leading the steam inlet pipe through a gland enables the correct nozzle position to be arrived at with little effort.

Blast pipe and blower

A too-large diameter blast nozzle will result in freedom from excessive back pressure but a soft exhaust and dull fire. The blast nozzle size can, of course, be calculated, but I have found it better to start with a small bore and increase if back pressure is found to be excessive.

As a guide, for a single cylinder engine 1⅜ inch bore by 2 inch stroke, a ⅛ inch bore nozzle gives a good sharp blast keeping an anthracite fire lively, but the ideal is to have a screw-in nozzle which can be changed if required, increasing the bore by 1/32 inch at a time.

The 3/16 inch dia nozzle originally fitted on the Durham engine proved too large for use with poor coal, while a ⅛ inch dia nozzle and firebars with at least ¼ inch gap between proved ideal.

The steam blower fitted on traction and other road engines, unlike that on a railway locomotive, did not deliver through a hollow stay.

To discourage its excessive use (and a heavier coal bill) the blower pipe ran from the cylinder casting to one side of the chimney base casting, with a small globe valve behind or beside the chimney. As the pipe in 2 or 3 inch scale need be only 1/16 inch bore, no nozzle is required, and the pipe should turn-up the chimney at about the same height as the blast nozzle.

Lubrication

Before the introduction of reliable mineral steam cylinder oil and grease, engines were lubricated by vegetable fats, rape seed oil, tar oil and animal fats. Goose grease and sheep fat were particular favourites and continued to be used for a surprisingly long time after higher boiler pressures and compound engines brought a speeding up in the development of mechanical lubricators. Even after the introduction of compounding for road engines in the 1880s, a suet lubricator, or ''fat cup'' was fitted by some makers to the low pressure cylinder, while a mechanical lubricator looked after the high pressure side.

To use the suet lubricator, the plug cock was closed, the screwed top removed, and the cup filled with inedible fat cuttings, pressed well down. After replacing the screwed top the cock was re-opened and the contents of the cup rendered slowly down by the steam, the accompanying smell varying according to the degree of rancidness of the contents. While not suggesting that this procedure should be followed nowadays, this type of lubricator, usually elegantly proportioned, is a useful addition in the event of mechanical lubricator failure, or for priming a cylinder prior to laying-up an engine. That good old standby, the displacement lubricator, was fitted by many makers as a standard fitting, even with mechanical pump lubrication, and if carefully proportioned need not look out of place in small scale. Whichever type of lubricator is fitted, oil should be led preferably into the valve chests from whence it will be carried through into both ends of the cylinder. It should never be fed into the top steam cavity or dome as surplus oil will inevitably find its way down the rising steam passage and into the boiler.

On a compound engine the HP exhaust will carry some oil through into the LP valve chest, but it is better to play safe and fit a teepiece in the oil delivery line with a pipe to both valve chests.

With a mechanical lubricator, a non-return valve will be required at the cylinder end of the delivery line to prevent steam entering the oil pipe; a displacement lubricator can of course be screwed straight into the valve chest.

The cylinder

Whether the cylinder casting has a curved base flange machined to sit directly on the boiler pad or a flat flanged base sitting on a flat topped saddle fixed to the boiler top, the joint between the two faces is of paramount importance as it must withstand boiler pressure.

Readers of my ''Countryman's Steam'' articles will probably have grown heartily sick of my stressing the need for the boiler to be clearly marked with the centreline along both top and bottom of the barrel, with the front end squared off truly at 90 degrees to the o/dia. Nevertheless, if this is done, fitting the cylinder is much easier, for the boiler front edge is the datum for the lining-up fixture and the cylinder bore is certain to end up parallel to the boiler centreline in both planes. Keep the holes in the base flange as close to setscrew size as possible to cut down the amount of possible movement of the casting on the boiler pad or saddle. Ideally the cylinder casting should be dowelled onto the boiler pad as this removes thrust from the setscrews themselves, but this is difficult unless the base flange extends well out from each end of the cylinder body. Again, a personal preference, but with a flat saddle mounting I turn a recess in the underside of the cylinder casting, around the bottom steam cavity, to take an O ring; when all setscrews are tightened down the ring should be compressed and the cylinder flange sitting down

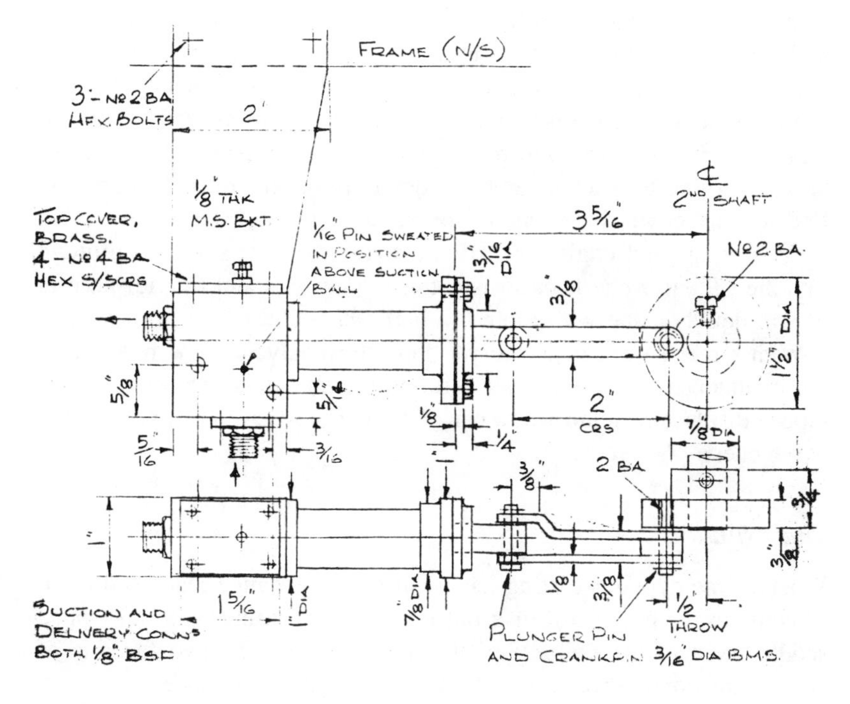

Fig. 20 Feed pumps

evenly all the way round. Smear all faces with Boss White or similar jointing, and make sure the O ring is the correct steam quality heat resistant variety, and the joint will last indefinitely.

With a curved base flange I restrict the jointing area to an oval ring of steam jointing cut to fit around the steam cavity and boiler steam holes, leaving the outer edges of the cylinder flange to pull down, nipping the jointing material well inside the outer edges of the flange – again all faces should be smeared with jointing compound, setscrews or studs should be a tightish screwed fit in the tapped holes, and as the holes will run through into the boiler, all should be treated to a dab of jointing compound too. Because it is impossible in small scale to reproduce the complex cored-steam jacketing and passages of the full-size cylinder casting, other ways of taking steam from the boiler and up into the top dome have to be used. A passage can be drilled to link the dome with the bottom steam cavity, but this will have to dodge steam and exhaust

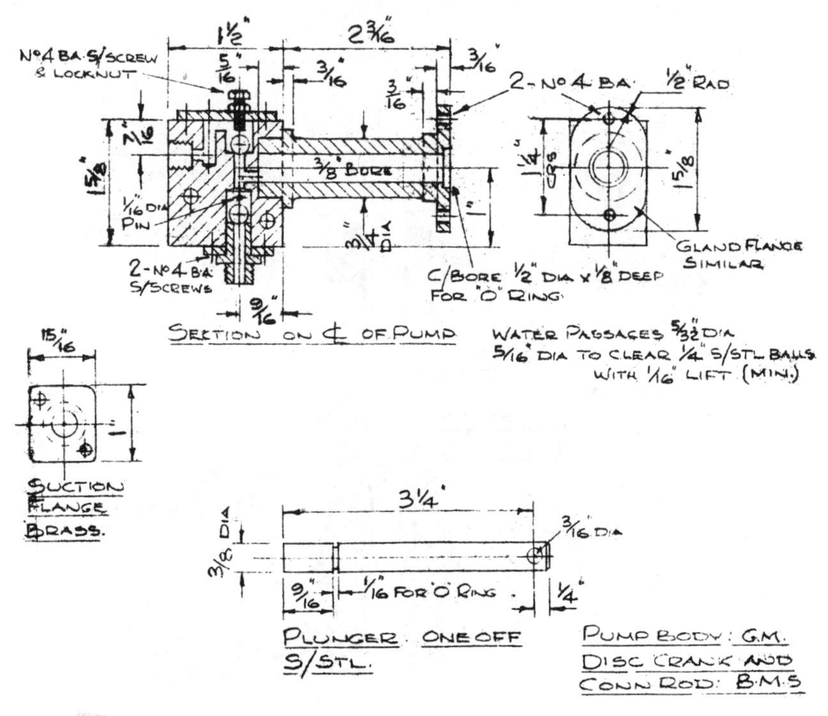

Fig. 21 Feed pumps

ports and contributes little towards efficiently steam jacketing the cylinder or cylinders.

On the other hand, if a cylinder liner is used, with an annulus machined midway along the bore into which the liner is to be pressed, steamways can easily be drilled connecting the annulus with both top and bottom steam spaces, and if made wide enough the annulus serves to jacket the liner and heat the whole casting.

To avoid running too close to the exhaust port, or requiring an excessive thickness of metal between port-face and cylinder liner hole, while at the same time allowing the maximum amount of steam into the annulus for jacketing purposes, I have introduced an eccentrically formed annulus. The cylinder of the 2 inch scale Fowler single-cylinder ploughing engine described in the *Model Engineer* has this feature which lends itself particularly well to the Fowler casting. (Fig. 23)

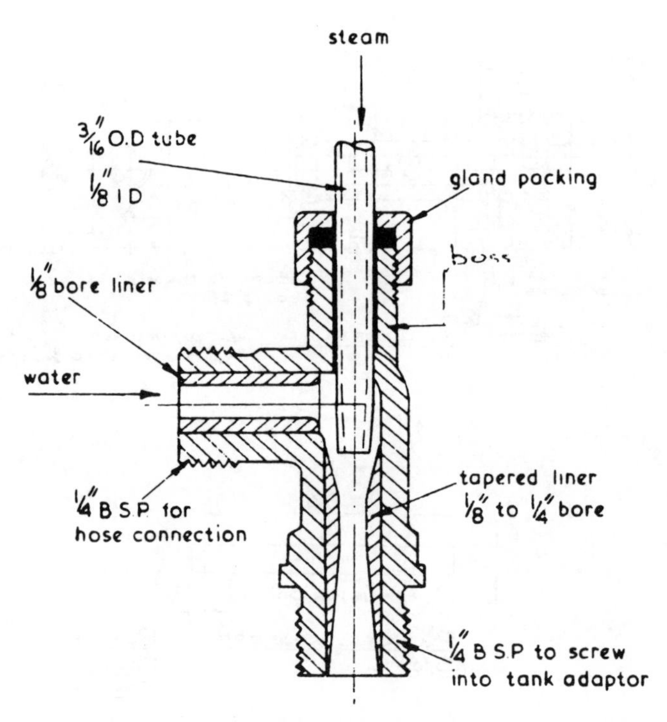

Fig. 22 Water lifter

The liner itself can be either phos. bronze or a good quality close-grained CI, but I have used, quite successfully, a steel one. Piston rings should be not less than two per piston, and not less than 1/16 inch square in section. In full size it was the practice to peg rings so that gaps were always opposite to each other, and despite what is sometimes said about never using similar metals running in contact with each other, in full size both piston rings and cylinder are cast iron – as of course they are in the majority of IC engines too, so disposing of that argument.

Still in full size, pistons were usually of wrought steel fitted on hammered-steel ground rods. In small scale gunmetal or bronze is suitable with a short length of counterbore to ensure concentricity on the rod.

Of recent years O rings have gained popularity in place of piston rings or graphited packing, but these must be the correct grade of material, fitted in correctly dimensioned grooves otherwise they are useless.

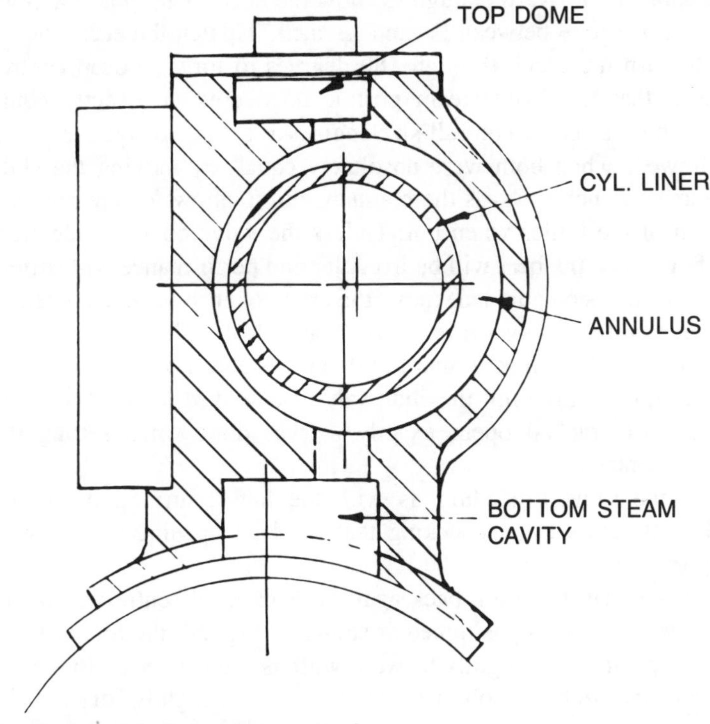

Fig. 23 Eccentrically bored steam annulus

Valve setting

The great majority of road and agricultural engines were fitted with Stephenson Link valve gear and slide valves. However, Fowlers also favoured Firth radial gear (derived from Hackworths) for their super-heated piston valve ploughing engines and certain of their rollers, and a few other makers used various forms of radial gear at times. Manns at one time even used a form of slip eccentric on their traction engines.

Whether the cylinder layout is single or compound, the same rules apply to setting the valves correctly, be the engine full size or small scale.

Taking a single-cylindered three-shaft engine as an example, on which the crank turns forward and the link is down when the engine is travelling forward, first turn the crank to back dead centre. Next, move the forward eccentric to a position exactly 90 degrees ahead of the crank (it will be found useful to scribe the centreline of each eccentric sheave on the outer face) and then very slightly turn it forward until the slide valve has moved forward enough to show the back steam port fractionally open for lead – between 1/64 and 1/32 inch. Tighten the eccentric on the shaft, turn the crank through 180 degrees to forward dead centre, and check that the front steam port is uncovered an exactly equal amount. This was known on full-size engines as setting to equal lead and was followed, when both were absolutely equal, by moving the slide valve about 1/16 inch towards the chimney end to allow for longitudinal expansion of the boiler when hot. Unless the valve opens to identical lead at both ends, the beat will be irregular and performance will suffer.

Once this has been attained, move the crank over to back dead-centre again and turn the backward eccentric, the rod of which is attached to the bottom eye of the link, round to 90 degrees ahead of the crank then advancing it carefully until the slide valve has moved forward to show the back port cracked open, exactly as was done when setting the forward eccentric.

Follow the same procedure as with the latter, turning the crank through 180 degrees and checking that the lead opening of the front steam port is equal too.

Now if we turn the crank back again to back dead-centre we should have the two eccentrics positioned as shown in Fig. 24, the forward one will point upwards and slightly forward with its rod connected to the top of the link, the backward one will hang down and slightly forward, its rod connected to the bottom of the link. The rods are thus in the "open rod'' arrangement.

If the engine is a four-shaft engine, as are all ploughing engines and quite a lot of tractions, the crankshaft will revolve backwards when the engine is moving forward. To save wear and tear on motion pins etc. it was customary to so arrange that the link was down when going forward and up when travelling backwards and to bring this about the rod from the backward eccentric has to be connected to the top of the link, the rod from the upper or forward eccentric connecting with the bottom of the link. This is known as the ''crossed rod'' arrangement. Most of my engines are designed with valves having no exhaust lap, the inside edges of the slide valve cavity lining up when in mid-position with the inside edges of the valve chest steam ports, known as ''line-in-line''. The outer edges of the slide valve, again in mid-position, extend over the outer edge of each steam port to give outside lap and a maximum port opening to steam of two-thirds of the port width. With the reversing lever in mid-gear position practically no steam should enter the cylinder.

Compound cylinders

Some years ago I contributed an article to *Model Engineer* on the historical and design aspects of the principle of compounding applied to road engines.

Without wearying you by repeating basic formulae relating to full-size engine design, there are several rules which must be observed if a compound cylinder layout is to work successfully in small scale. Obviously the compound traction or ploughing engine requires a higher boiler pressure to ensure that the exhaust from the low pressure cylinder into the blast pipe carries sufficient pressure, after expansion in the cylinder, to provide an adequate blast in the chimney.

For this reason, and to avoid raising boiler pressure more than necessary, I used a slightly lower cylinder ratio of just under 2:1 when designing the Fowler Class BB/1 in 2 inch scale.

This is well under the usual ratio of LP to HP cylinder area of between 2¼:1 and 3:1, and it ensures that even at boiler pressures, well under 100 psi the blast keeps the fire lively.

The first rule is then to remember that in referring to cylinder ratios, it is the respective cylinder areas which are used in the small calculation and not the diameters, and that the ratio between LP and HP area must be kept lower than in full size.

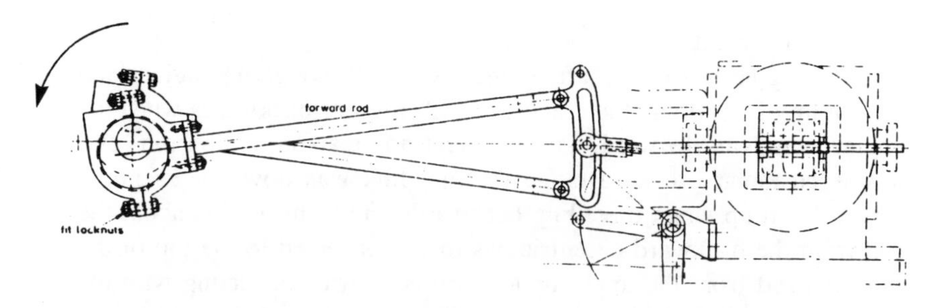

Fig. 24A Arrangement of rods — crossed

The second rule is that the receiver volume, that is the space occupied by the exhaust steam between leaving the HP cylinder and arriving in the LP must be not so large as to absorb all expansive effort before the work is done on the LP piston, a critical point and particularly in small scale. The shorter the passage between HP exhaust port and LP valve chest, the better.

Boiler lagging

To conform to full-size practice the boiler should be lagged with narrow strips of wood laid lengthways and held in place by wire bands. In small scale it is easier to use 1/16 inch thick asbestos millboard, taped in place.

Boiler cladding, like the lagging material, has to be cut to fit around cylinder base flange and valve spindle guide bracket, if one is fitted. The most convenient way of doing this is to fit both in two separate halves, first tinning along both top edges of the cladding material, usually brass or tinplate. Overlap the cladding about 1/8 inch on top of the boiler and allow a slightly longer overlap along the bottom seam, securing both

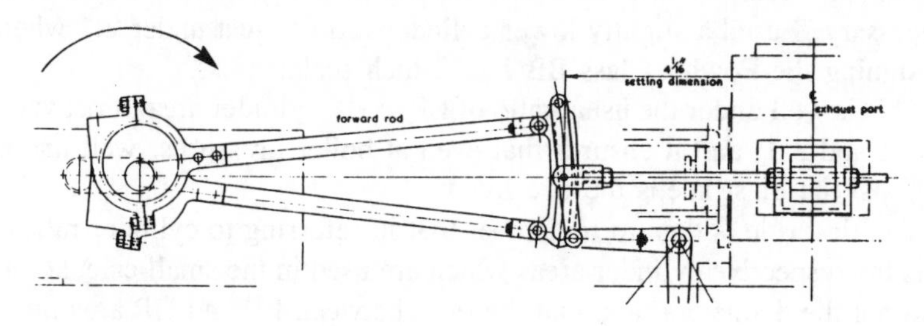

Fig. 24B Arrangement of rods — open

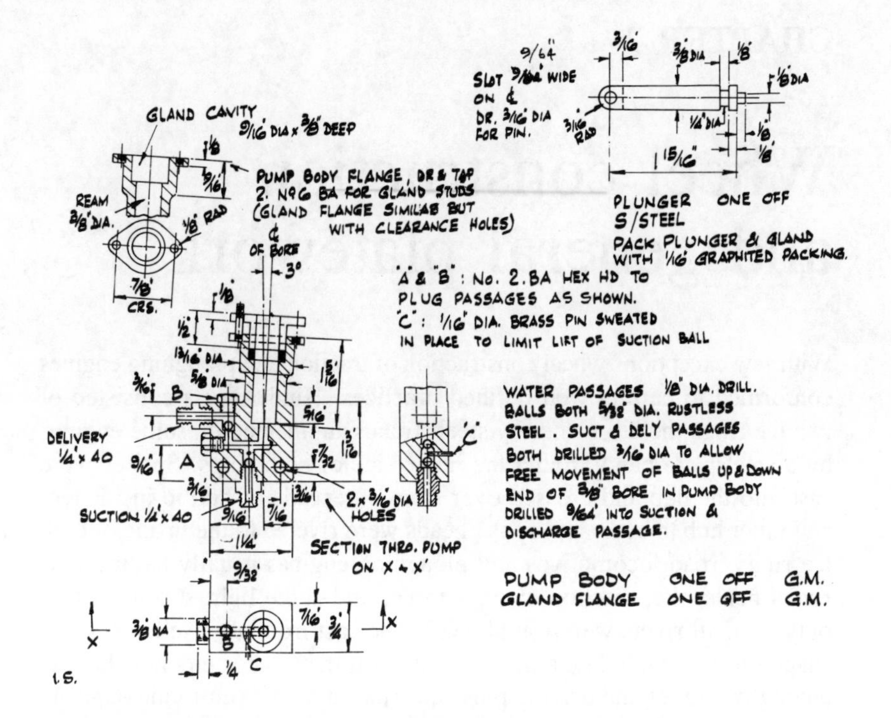

Fig. 25 Feed pump

halves in position with thin wire, and running a soldering-iron along the top overlap to form a continuous soldered joint.

Replace the holding wires by boiler bands and tighten to close the lower overlapping seam.

The position of clack-valve holes etc, in each half of the cladding sheet can be marked and cut from a paper template before bending around the boiler barrel.

Recommended for boiler insulation in place of asbestos millboard is *kaowool* ceramic fibre blanket ¼″ thick, compressing to ⅛″ thick.

CHAPTER 7

Wheel construction and general platework

With few exceptions wheel construction of traction and ploughing engines conformed to certain well defined practices. Hind wheels consisted of two tee rings, held together by steel strakes augmented on some engines by a steel plate band around the inner face of both rings. Spokes were cast into the hub and crossed over from outer side of hub to inside tee, and inner hub to outer tee; spoke heads were riveted to the inside of both tee rings, road locomotives and ploughing engines usually having four rivets per spoke, traction engines three, and some light steam tractors only two; all rivets were round-head. The cast iron hub itself was pear-shaped to accommodate a driving pin – on many later engines the hub having two lobes and driving pins equi-spaced on the centreline opposite each other. The outside face of the hub was set back from the edge of the wheel, so that spokes were of differing lengths each side of the wheel (see Fig. 26). Strakes, riveted to both tee rings with counter-sunk rivets, were set at an angle to the horizontal and when viewed from behind the engine, inclined upwards towards the inner edge of the wheel, although this was not an invariable rule as several early Fowler ploughing engines and some less well-known traction engines had strakes inclining the opposite way.

Generally speaking the strake angle was steeper on early engines, flattening out on those built from the 1890s onwards from 27 degrees or so to around 15 degrees from the horizontal.

In an attempt to limit damage to road surfaces the law stipulated that the distance between strakes must not exceed 3 inches and the width be not less than 3 inches, with a maximum thickness of ¾ inch. Commonly used were strakes of 3 to 4½ inches width and ½ inch thickness, spaced so that the end of one overlapped the opposite end of the next to give continuous contact with the ground or road surface. Outer edges of both tee rings were usually curved slightly downwards, giving a radiused effect and strakes, as well as being bent to follow this form, also stopped just short of the outer edges of both tee rings to avoid sharp edges projecting when road wear and crushing had hammered the metal thin at the edges.

Front wheels were built up in the same way as hind wheels, with steel spokes cast into the hub and riveted to alternate sides of a single tee ring, the exceptions being the majority of ploughing engines and some crane engines which had extra wide wheels made up of two tee rings. Rivets were usually three per spoke, reduced to two on some steam tractors. Solid tyres on steel rings were bolted over the strakes of heavy haulage and showmens road locomotives until, in the 1920s it became the practice to press steelbacked solid tyres directly on to the wheel, omitting strakes altogether. An act of Parliament passed some years earlier, required all engines except "locomotive ploughing engines" to be mounted on rubbers by the close of 1939. One happier result of the war was the postponement (official or otherwise) of this requirement, many engines engaged on threshing work remaining on their original steel strakes until the coming of the combine in the early 1950s.

Turning from full-size practice to 2 inch scale, two schools of thought prevail on the best method of attaching spokes to hub on both hind and front wheels. The first is to mill separate slots each side of the hub for spokes, the outer face of each spoke finishing up level with the outside of the hub, retained in place by an outer plate attached to the hub by countersunk-head screws. The second method is to turn down each end of the hub to form a protruding spigot equal in length to the spoke thickness, plus the thickness of an outer plate 3/32 to 1/8 inch thick, attaching each spoke to the hub flange with a 4 BA countersunk-head steel screw and pressing on to the remaining short length of spigot the outer plate. This should be shaped to the same contour of the hub itself for hind, and circular for front wheels. These plates can be secured in place either by three c/sk head screws or by utilising three of the spoke screws. This method of construction gives a very strong wheel without the task of milling separate slots for each spoke into the hub, but on the debit side it does leave a gap between each spoke which then requires filling. On earlier engines I laboriously cut slots in a circular plate into which the spokes were inserted, but for the Aveling and Porter roller and every engine since, have used a metallic filler, firmly pressed into the gaps between the spokes and filed to bring flush with the surrounding metal. The filler I have found to be most suitable is Belzona Molecular

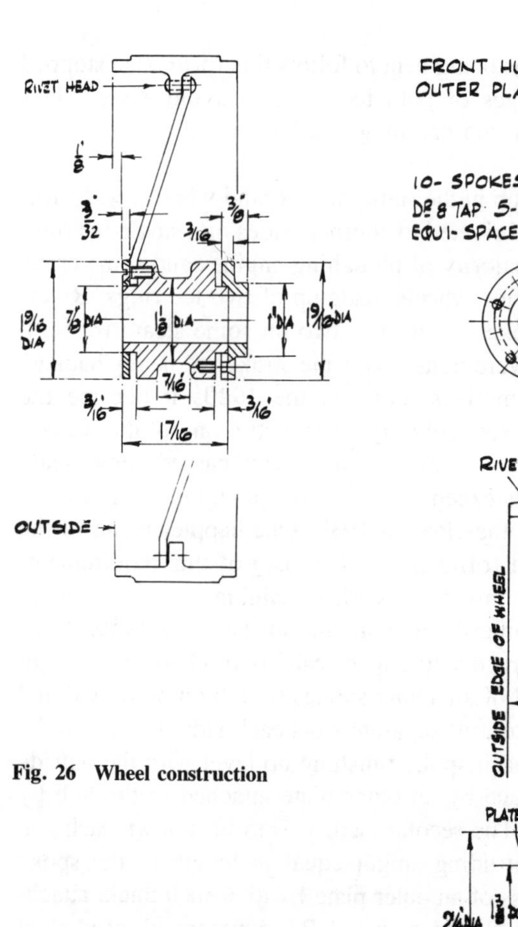

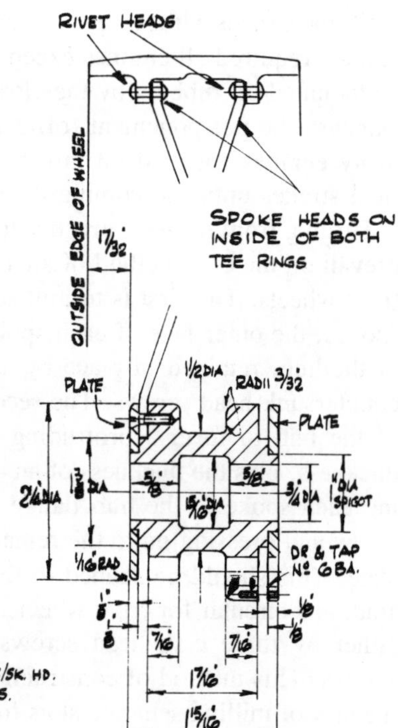

Fig. 26 Wheel construction

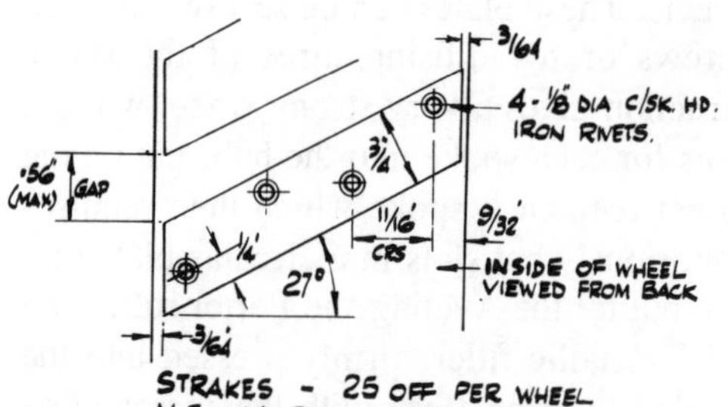

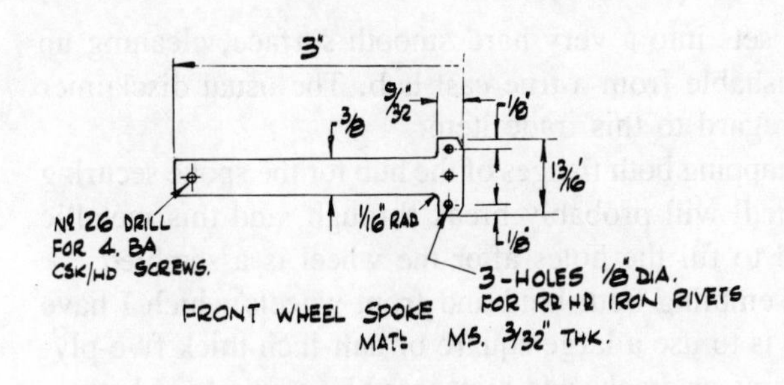
3"
9/32
1/8
3/8
13/16
Nº 26 DRILL
FOR 4. BA
CSK/HD SCREWS.
1/16" RAD
1/8
3-HOLES 1/8 DIA.
FOR RD HD IRON RIVETS
FRONT WHEEL SPOKE
MAT'L M.S. 3/32" THK.

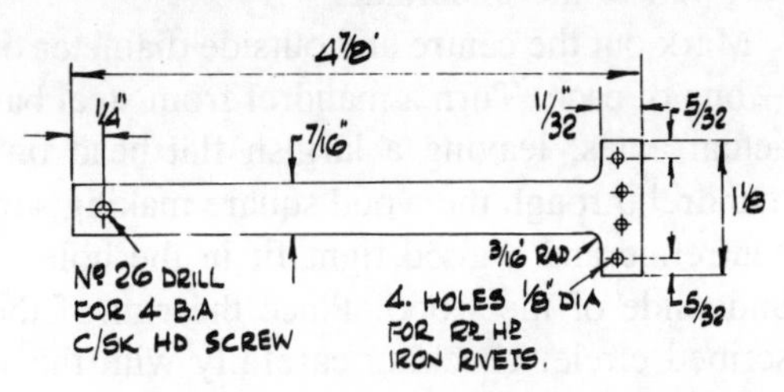
4 7/8"
1/4"
11/32
5/32
7/16"
1 1/8
3/16 RAD
Nº 26 DRILL
FOR 4 B.A
C/SK HD SCREW
4. HOLES 1/8" DIA
FOR RD HD
IRON RIVETS.
5/32
HIND WHEEL SPOKE 14-OFF PER WHEEL
MAT'L M.S - 1/8" THK.

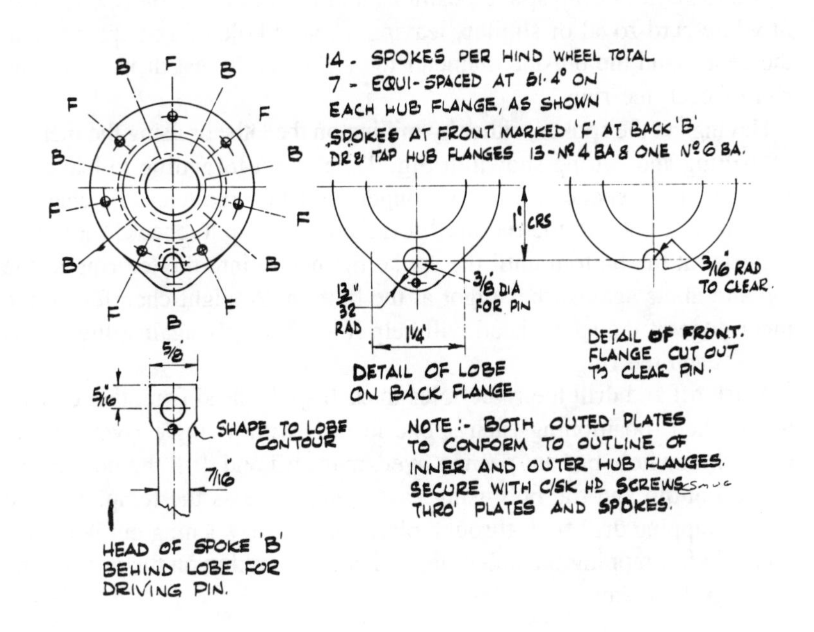
14 - SPOKES PER HIND WHEEL TOTAL
7 - EQUI-SPACED AT 51·4° ON
EACH HUB FLANGE, AS SHOWN
SPOKES AT FRONT MARKED 'F' AT BACK 'B'
DR & TAP HUB FLANGES 13 - Nº 4 BA & ONE Nº 6 BA.
B
F
B
F
F
B
B
F
F
B
B
F
F
B
1" CRS
3/8 DIA
FOR PIN
13/32" RAD
1 1/4
DETAIL OF LOBE
ON BACK FLANGE
3/16 RAD
TO CLEAR.
DETAIL OF FRONT.
FLANGE CUT OUT
TO CLEAR PIN.
5/8
5/16
SHAPE TO LOBE
CONTOUR
7/16
HEAD OF SPOKE 'B'
BEHIND LOBE FOR
DRIVING PIN.
NOTE :- BOTH OUTED' PLATES
TO CONFORM TO OUTLINE OF
INNER AND OUTER HUB FLANGES.
SECURE WITH C/SK HD SCREWS
THRO' PLATES AND SPOKES.

Super Metal, which sets into a very hard smooth surface, cleaning up to appear indistinguishable from a true cast hub. The usual disclaimer is added here with regard to this trade item.

When drilling and tapping both flanges of the hub for the spoke securing screws the tapping drill will probably break through, and this metallic filler should be used to fill the holes after the wheel is assembled.

The method of assembling both hind and front wheels which I have found to be handiest is to use a large square of half-inch thick five-ply, flat and free from warps or cracks and reinforced by a couple of battens screwed to the underside.

Mark out the centre and outside diameter of the back and front wheels – one of each. Turn a mandrel from steel bar, the same diameter as the actual axles, leaving a largish flat head on the lower end. Insert the mandrel through the wood square making sure it is exactly on the wheel centreline and a good tight fit in the hole, with head hard against the underside of the wood. Place the rim of the wheel in position on the scribed circle, checking carefully with the vertical mandrel projecting through the wood for concentricity, and then screwing in position three small wooden blocks at 120 degrees to each other and hard up tight against the outside diameter of the rim (tee ring).

Next, mark out the spoke positions and rivet holes radially on a disc of white card-royal or similar, leaving a round hole of axle diameter in the centre and the outside diameter the right size to just drop inside the rim of each tee ring.

Having cut the spokes, either by milling on the lathe or using the method of drilling and cutting shown in Fig. 27, and not forgetting to leave the first couple of spokes a fraction longer until the true length is checked "to place", bend to the required form and offer up to check, removing material at the bottom until the spoke fits neatly into the tee ring at the top and abuts against the spigot at the bottom. A slight chamfer on the inner edge of the spoke head will help it to sit snugly against the radius inside each tee ring.

Mark off and drill the rivet holes in each spoke head from the tee ring holes, then, positioning each spoke in turn with a single rivet pushed through one hole but not peened over, mark off and drill the hole in the spoke bottom and the hub using each outer plate as the template. Use a 4 BA tapping drill first, through plate and spoke, opening out with No. 26 drill after tapping the hub hole, and countersinking on the outside the holes in the plates.

Don't forget that one spoke at the back has a broad base to allow the driving pin to pass through, and is the only one in each wheel to be secured in place by a 6 BA and not a 4 BA. The spokes are marked, on my drawing, viewed from the outside of the wheel, B for back of hub, F for front of hub. Also don't forget a generous radius (minimum 1/16 inch) on the outside of the outer plates to simulate the appearance of a casting as far as possible.

I deal with three spokes on each side first, checking the wheel for eccentricity after riveting and screwing these in position, before going on to the remainder. Drill through 3/32 inch dia from the centre portion of all four hubs, into the relief or the bore on front hubs, for the grease cap tube.

Using the above method of assembly, right at the start, when positioning the hub on its mandrel, pack up on flat washers to obtain the correct relative position from face of hub to outer edge of wheel.

The wheels described in the foregoing constructional notes are those for the 2 inch scale Durham and North Yorkshire traction engine, and it should be noted that the rear hub differs slightly from other makers in having a full driving pin lobe on the inside only – a practice only followed by one or two other firms.

The tee rings for this engine were turned from heavy gauge steel tube, enabling both hind wheel tees to be turned from one single length with a shallow groove on the centreline simulating the visible joint between the two rings.

The tube material used is thick enough to allow the tyre on the front wheel to be turned integral with the tee ring. Tyres were riveted in place, usually in two or more lengths with an angled joint between, and this should be represented by a saw-cut at 45 degrees across the upstanding tyre width.

When turning the combined width of two tee rings from a single length of thick-walled tube it is advisable to leave the central section, between the tee legs, slightly over-scale in thickness, to prevent distortion during machining. Rolling the rim from 6 swg (.192 inch) thick ms sheet and welding or brazing in a flat ring produces a strong wheel; the double tee rings of hind wheels can be rolled as a single length, with both legs turned to size and welded in, and providing the rolling operation has produced a truly circular rim, will need very little cleaning up afterwards.

Taking the Durham and North Yorkshire wheels as an example, the legs of all tee rings are 3/16 inch thick and care is necessary to prevent

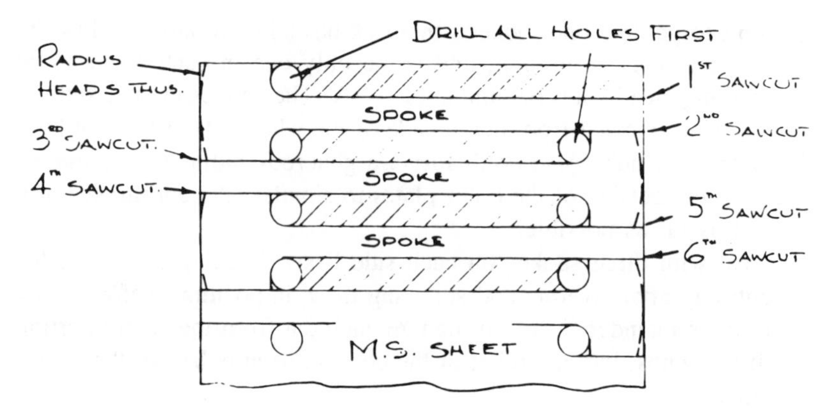

Fig. 27 Method of cutting spokes

them distorting badly when heated – particularly if it is not intended to machine the inside of the tee rings after fabrication.

I think the best way to minimise the risk of distortion is to set the ring up as shown in Fig. 28 with the flat ring firmly secured by at least four clamping pieces – I use six – equi-spaced around the inside diameter and at the same time supporting the outside of the rim with ms stops and clamping the top edge with another set of six equi-spaced studs and clamps.

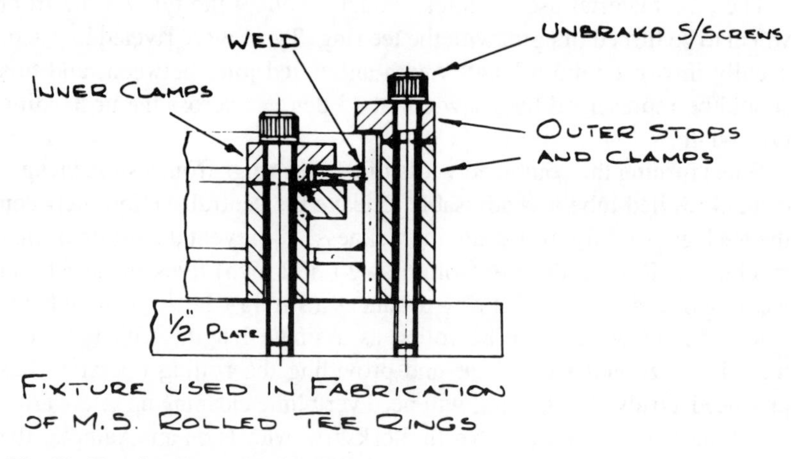

Fig. 28 Tee ring fixture for fabrication

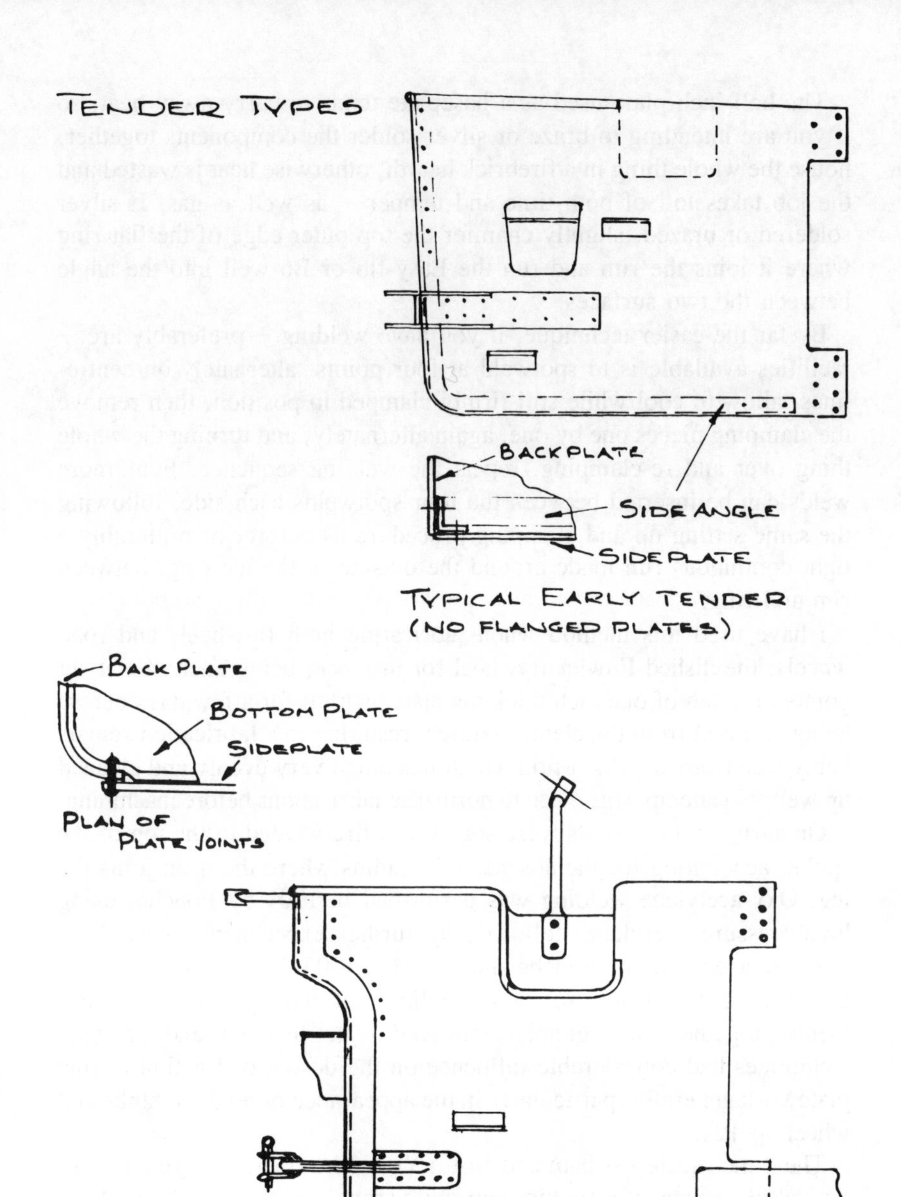

Fig. 29 Tender construction

The half-inch plate used as a baseplate tends to carry away heat, so if you are intending to braze or silver solder the components together, house the whole thing in a firebrick hearth, otherwise heat is wasted and the job takes toll of both time and temper – as well as gas. If silver soldered or brazed, slightly chamfer the top outer edge of the flat ring where it joins the rim and run the Easy-flo or B6 well into the angle between the two surfaces.

By far the easier technique, if you have welding – preferably arc – facilities available is to spotweld at four points, alternately on centre-lines, allow to cool while still firmly clamped in position, then remove the clamping pieces one by one, again alternately, and turning the whole thing over and re-clamping, repeat the welding sequence. Four more welds can be inserted between the four spotwelds each side, following the same setting up and clamping procedure as before, or preferably a light continuous run made around the outside of the tee ring, between rim and ring.

I have used this method when fabricating both flywheels and road wheels, the dished Fowler flywheel for instance, being anchored at six points to a slab of one inch thick ms plate and left for a few days before being released from the clamp fixtures, resulting in a fabrication remarkably free from any distortion which machined very evenly and cleaned up well. No attempt was made to normalise fabrications before machining.

On early engines heads were sometimes fire-welded to the top of the spoke, accounting for the absence of a radius where the head joins the leg. Oxy-acetylene welding was performed in 1901 by Fouche, using high-pressure acetylene, followed by further experiments using low-pressure acetylene, by Fouche and Picard in 1903, leading to its general use from then onwards for both welding and cutting. These developments, together with greatly improved plate forming and pressing techniques had considerable influence on the design of traction engine platework generally, particularly in the appearance of tenders, tanks and wheel spokes.

The 2 inch scale Durham and North Yorkshire traction engine is very typical of engines of a century ago, with straight-backed tender built up on an angle-iron frame, spokes with minimal head and radius, flat plate smokebox door on a very short smokebox, and indeed, no platework frills at all. Compare these characteristics with a Fowler engine built from about 1911 onwards, when a tender was introduced with straight sideplates riveted to deeply flanged and radiused back plate and sloping bottom plate. (Fig. 29)

The two inch scale class BB ploughing engine typifies this strongly constructed and very handsome back-end, together with the generously radiused spoke heads fitted to the wheels of Fowler, and, by this time, most other engines.

As an alternative method of wheel building, the spokes can be cut from two circular plates in the form of a spider, leaving the centre as a circular disc to fit over each hub spigot; in fact this form of construction was employed on quite a number of full-size engines both here and overseas.

Again, marking out each spider will be far easier if it is drawn full size on card, which can then be cut and used as a template.

Unfortunately, nothing detracts more from the appearance of an engine than faulty wheels; the most difficult error to disguise is the use of excessively deep and unrealistic tee-rings, sometimes coupled to a too-narrow wheel. All traction engine makers used tee rings of just sufficient depth to take the spoke heads, commensurate with adequate cross-sectional area for the loading imposed. Any extra length in the leg of the tee section was wasted material, and in any case would have made rolling the section difficult. If alternative wheel widths were offered by an engine manufacturer, as they often were, I always think it wiser from the appearance point of view to use the wider when building in 2 and 3 inch scale, and to avoid strakes of excessive thickness, which should be left unpainted if a realistic finish is wanted. Even on a new engine, in full size, the area between each strake very quickly lost its pristine appearance and became rusty!

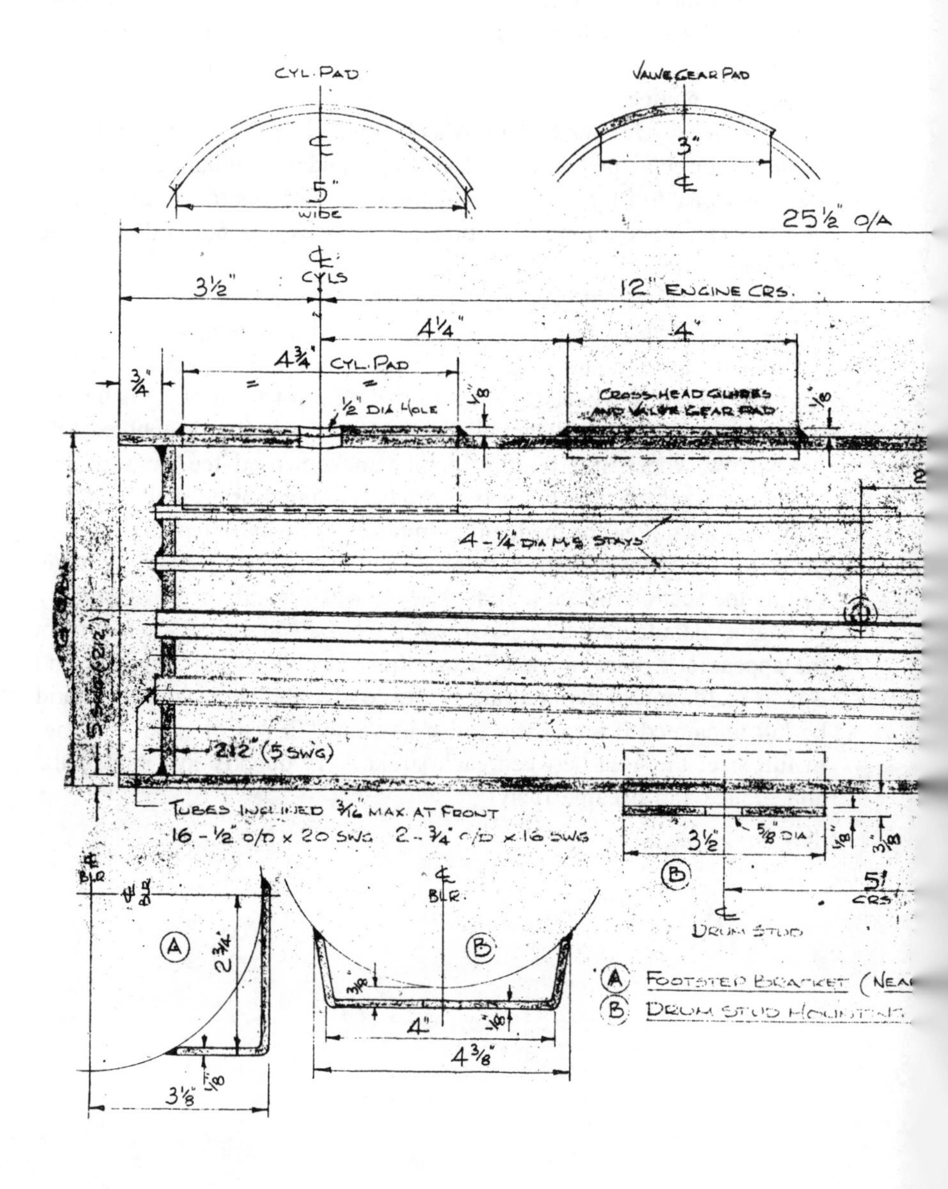

Fig. 30 Fowler Superba class ploughing engine in 2″ scale boiler (steel)

BOILER [illegible] M.S. THROUGHOUT, THE SHELL
OF [illegible] DIA HOT-FINISH SEAMLESS TUBE, 5 SWG (·212") THK
WITH PLATE THICKNESS NOT LESS THAN SPECIFIED, FOR ALL PLATEWORK.
ALL WELDS CONTINUOUS SEAM HEAVY FILLET FULL STRENGTH ELECTRIC ARC.
WORKING PRESSURE: 100 P.S.I. HYDR. TEST PRESSURE: 200 P.S.I.

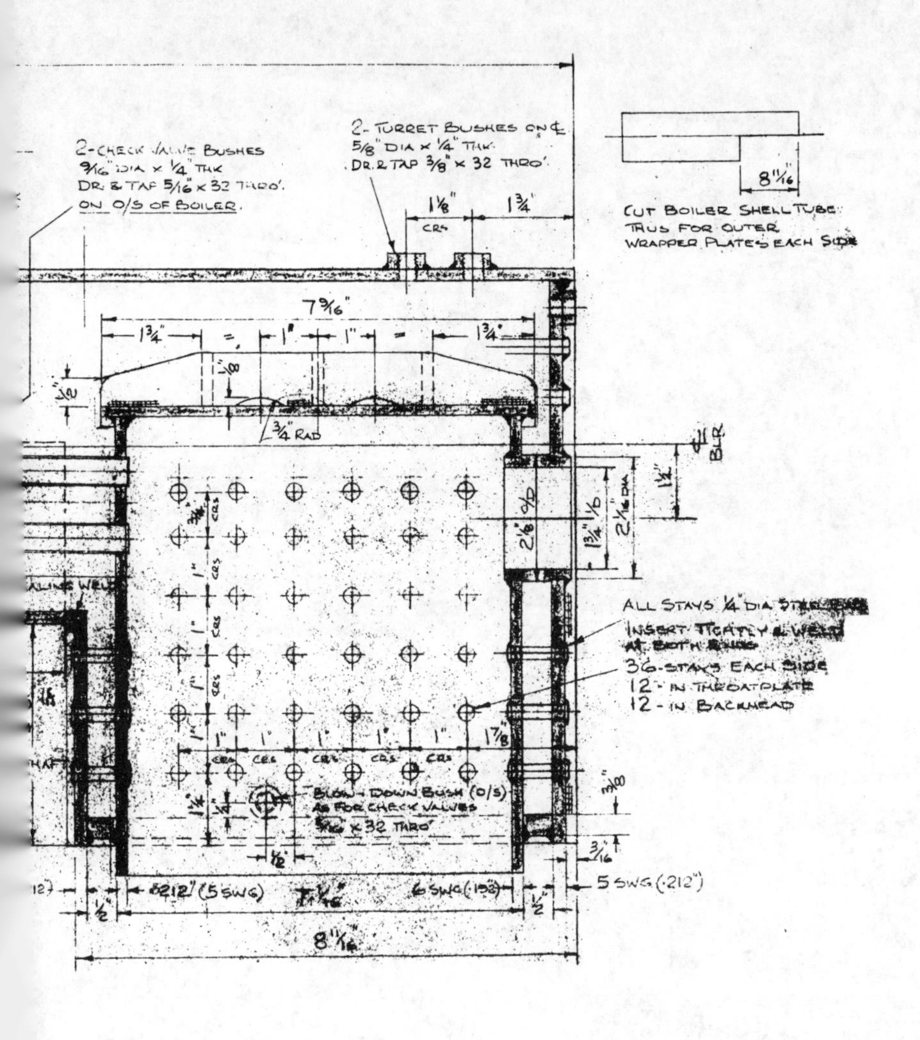